Physics and Chemistry in Space
Volume 4

Edited by
J. G. Roederer, Denver, and J. Zähringer, Heidelberg

A. Omholt

The Optical Aurora

with 54 figures

Springer-Verlag New York Heidelberg Berlin 1971

A. Omholt

Universitet i Oslo, Fysisk Institutt, Blindern
Oslo, Norway

ISBN 0-387-05486-3 Springer-Verlag New York Heidelberg Berlin
ISBN 3-540-05486-3 Springer-Verlag Berlin Heidelberg New York

Acknowledgements

The author is grateful to a number of his colleagues at the Norwegian Institute of Cosmic Physics, Blindern, and the Auroral Observatory, Tromsø, for reading parts of the first draft and offering helpful comments: C.S. Deehr*, A. Egeland, O. Harang, A. Haug, O. Holt, G.J. Kvifte, and H. Pettersen. Particular thanks are due to G.T. Hicks**, who read both the first and second drafts and who also offered most helpful advice on the English.

It is also a pleasure to acknowledge the able work done by the photographic laboratory and drawing office of the University of Oslo, and by Mrs. Gerd Solberg on the references. Finally, it is a pleasure to thank Mrs. Anna-Sophie Andresen for her patient and excellent work with the manuscript.

* On leave from Geophysical Institute, College, Alaska.

** On leave from Naval Research Laboratory, Washington D.C.

Preface

The aim of this book is to describe and discuss the aurora as an optical phenomenon, one which can be observed by the naked eye as well as with more sensitive optical detectors. It continues the tradition of studying that impressive and imaginative play of nature, the northern lights, seen and discussed by the Greek philosphers as early as the sixth century B.C.

Today the study of the optical aurora is only one of many ways of acquiring information about a major phenomenon: the ejection of plasma from the sun, the interaction of this plasma with the geomagnetic field and the injection of fast particles into the earth's atmosphere. Hence, the separate treatment of the optical aurora is justified by the particular scientific approach: detection and interpretation of electromagnetic radiation, approximately in the 1000–100000 Å region, produced through interaction between the auroral particles and the earth's atmosphere.

Other techniques, such as radio observations, X-ray observations, direct particle detections from rockets and satellites, studies of magnetic storms, and measurements of the magnetic field and plasma properties in the magnetosphere, are as important or more important than the classical way of studying the optical aurora. Nevertheless, it was felt worthwhile to treat the optical aurora in a separate book, perhaps mainly because today one author cannot master the whole subject with sufficient competence. This book is thus one volume in a series of books giving a more complete picture of physics and chemistry in space.

The study of the optical aurora has two distinctly different major purposes; first, to consider the optical radiation as one of several end results of a vast phenomenon extending between the sun and the earth. In this connection the earth's atmosphere may be considered as a scintillating screen, and the purpose of the study of auroras is to acquire information about the particle streams penetrating into the atmosphere and the fundamental processes which produced them. Second, to use the information concerning optical radiation to deduce the effects of primary particles on the ionosphere, and hence to study the perturbations of the ionosphere caused by these particles. These studies are important in regard to the propagation of radio waves and other radio phenomena.

Both aspects will receive due consideration in this book, although it is neither possible nor appropriate to keep them separate, chapter by chapter, or to present the book in two separate parts.

Many of the subjects treated in this book have been comprehensively dealt with in an excellent book by J. W. Chamberlain (1961)*. However, since 1960, when that book was written, many important publications have appeared, and the state of our knowledge has advanced considerably. Here we shall emphasize the progress made since then, rather than repeat general descriptions and knowledge which was available at that time. Hence the reader will, for many detailed discussions, be referred to that book.

No historical treatment will be given. Brief summaries of the history of the study of the aurora have been published by Chapman (1967, 1969). Further information may be found in the above mentioned book by Chamberlain (1961).

Blindern, Norway
June 1970

Anders Omholt

* For references, see under Chapter 1.

Contents

Chapter 1

The Occurrence and Cause of Auroras: a Short Introduction

Chapter 2

The Electron Aurora: Main Characteristics and Luminosity

Chapter 3

The Proton Aurora

Chapter 4

The Optical Spectrum of Aurora

Chapter 5

Physics of the Optical Emissions

Chapter 6

Temperature Determinations from Auroral Emissions

Chapter 7

Pulsing Aurora

Chapter 8

Optical Aurora and Radio Observations

Chapter 9

Auroral X-Rays

Frequently used Symbols

Standard spectroscopic nomenclature is used throughout.
Some symbols are occasionally used for other quantities, in agreement with current literature. This use should be clear from the text.

α	recombination coefficient
A_{nm}	transition probability
$\boldsymbol{B}, B$	magnetic field (vector, scalar)
e	electronic charge
E	specific particle energy
$\boldsymbol{E}, E$	electric field (vector, scalar)
Φ	geomagnetic latitude
H	scale height
$H\alpha$, $H\beta$	Balmer alpha, beta lines
h	height
I	surface brightness, integral invariant
$4\pi I$	Auroral brightness
L	McIlwain invariant parameter
Λ_{L}	invariant geomagnetic latitude
λ	wavelength
m	particle mass
μ	magnetic moment
ν	frequency, collision frequncy
Ω	solid angle
ω	angular frequency, gyro frequency
Ψ	wave function
R	Rayleigh (photometric unit)
R, r	radius, space coordinate
σ	cross section (atomic or molecular)
t	time
T	universal time
τ	lifetime
Θ	pitch angle
$\boldsymbol{v}, v$	particle velocity (vector, scalar)
v', v''	vibrational quantum number
W	kinetic energy
ξ	penetrated air mass (STP)
ζ	penetrated air mass in number of particles

Chapter 1

The Occurrence and Cause of Auroras: a Short Introduction

1.1 Local Auroral Forms

1.1.1 Geometry

It is probably well known to the reader that the Aurora Borealis, in the north, and the Aurora Australis, in the south, appear as faint, luminous phenomena during the night. They appear most frequently in two zones around the magnetic poles, the auroral zones. Auroras occur, however, more or less over the entire polar caps and also to a certain extent equatorwards of the auroral zones. The appearance of auroras is connected with disturbances in the geomagnetic field, the aurora moving equatorwards with increasing magnetic disturbance.

Looking at a typical auroral display, one sees a bewildering number of auroral forms and situations, However, if analyzed in detail, each instantaneous auroral situation may be considered to be composed of various superimposed elementary auroral forms or structures, each varying in space and time. For practical purposes, we may limit ourselves to four such elementary forms or structures: (1) The quiet homogeneous arc or band streching across the sky in a straight or curved line; (2) auroral rays, which may vary considerably in length; (3) diffuse and irregular auroral patches; (4) large homogeneous surfaces. Their intensities may vary over several orders of magnitude. The elementary auroral forms (1) to (3) are illustrated in Figs. 1.1 to 1.6 (see next pages).

The quiet homogeneous arc is usually elongated approximately in the geomagnetic east-west direction, except in the central polar regions, where the orientation is approximately towards the sun. Homogeneous arcs more than 1000 km long have been reported (Akasofu, 1963, Feldstein, Khorosheva and Lebedinsky 1962). The width of an arc may vary from less than 1 km to several tens of kilometers (cf. Sect. 2.6). A generalization of the homogeneous arc is the homogeneous band. This

Fig. 1.1 Auroral arc (Photo: S. Berger)

Fig. 1.2 Homogeneous arcs and bands (Photo: S. Berger)

Fig. 1.3 Homogeneous band (Photo: S. Berger)

Fig. 1.4 Rays (Photo: S. Berger)

Fig. 1.5 Diffuse, patchy aurora (Photo: S. Berger)

Fig. 1.6 Auroral drapery, bands with ray structure (Photo: S. Berger)

form is also much elongated, but does not have the regular shape of an arc. It is usually folded into S-forms or spiral-like forms. Arcs and bands usually lie at heights between 100 and 150 km.

Auroral rays may be seen embedded in arcs or bands, or as more isolated structures. Their length may vary from a few tens to several hundred kilometers, stretching from 100—150 km and upwards. They are always elongated along the magnetic field lines, with horizontal dimensions from a few tens of meters to several kilometers.

Irregular, diffuse patches each generally cover an area of the order of magnitude of 100 km^2 or so; usually many such patches occur simultaneously. The large homogeneous surfaces, on the other hand, cover much of the visible sky with fairly constant and weak luminosity.

During an auroral display several, or all, of these forms may appear simultaneously, partly overlapping or embedded in each other. Auroral arcs and bands thus frequently show ray structure. Each form usually moves and varies in intensity with time. For homogeneous, quiet forms the velocities are small, up to a few hundred m s^{-1}, whereas particular forms with ray structure may show rapid motion, velocities of up to 50 km s^{-1} having been measured for individual rays (cf. Sect. 2.6). Simultaneously there are intensity variations in the moving forms. Large-scale intensity variations are combined with movements and variations in the gross pattern of the aurora and take place over minutes or hours. Other intensity variations are more rapid. In homogeneous forms intensity pulsations with frequencies up to about 10 Hz may occur, although intensity variations with a period of a few seconds are more frequent (cf. Chapt. 6).

The *International Auroral Atlas* (1963) gives a very good review of the common auroral forms and their classification. We shall return to a more detailed description and interpretation of characteristic properties of aurora in the following chapters; the common auroral display is described in some detail in Sect. 1.1.3.

1.1.2 Intensity

The aurora appears from a distance as a luminous cloud, having a certain surface brightness. For visible light, however, there is negligible self-absorption within the luminous regions, so that the apparent surface brightness is proportional to the thickness of the luminous aurora, or, more correctly, proportional to the emission per unit volume integrated along the line of sight. Since, in most cases, only the apparent surface brightness can be measured, this is used to define the intensity of the aurora, although for physical purposes the emission per unit volume is of considerably more interest.

If I is the surface brightness of an aurora observed along a particular direction, then $4\pi I$ is the total emission from a column of unit area within the aurora aligned along the direction of observation. This total emission is usually measured in photons $cm^{-2}\,s^{-1}$. To indicate that it is not a true surface brightness but emission from an emitting column, the units are often, though somewhat inaccurately, written as photons cm^{-2} (column^{-1}) s^{-1}. A more convenient unit for $4\pi I$ is the Rayleigh, abbreviated R. One R is equal to a total emission of 10^6 photons cm^{-2} (column^{-1}) s^{-1} (cf. Chamberlain 1961, Appendix II).

It is important to understand that the intensity of a particular auroral form depends on the direction from which it is observed. A thin luminous layer covering a large part of the sky is much more intense when viewed laterally at low elevation angles than from underneath.

Since the auroral spectrum consists of particular emission lines and bands, the intensity of an aurora is usually not referred to the total emission, which is difficult to measure, but rather to the emission in one particular line or band. A standard reference very often used is the strong green oxygen line at 5577 Å, which is one among several lines and bands which dominate the visual intensity. A convenient logarithmic classification system of auroral intensities is the "International Brightness Coefficient", which is defined in such a way that auroras of international brightness coefficients I, II, III, and IV emit the green oxygen line with an intensity $4\pi I$ equal to 1, 10, 100, and 1000 kR (1 kR $= 10^3$ R), respectively.

The intensity distribution is usually fairly constant among the most prominent auroral emissions within the range of wavelengths to which the human eye is sensitive, so that the brightness coefficient as defined above is usually proportional to the visual intensity observed by the eye. There are exceptions, however, such as auroras showing a strong red lower border, or diffuse, widespread red auroras, for which the brightness coefficient defined through the green oxygen line may not correspond to the visual intensity, because the ratio of other major visible emissions to the green line is changed. Another source of error in visual assignments of a brightness coefficient arises when the sky is covered with a widespread glow, so that no reference point is available to the observer.

Typical green-line intensities of diffuse, patchy auroras range from below the threshold of visibility, which is at approximately 1 kR, upwards to a few kR. Homogeneous arcs are usually of moderate intensity, from about one to a few tens of kR. The greatest intensities occur during short periods, outbursts, with vivid motions of rayed bands. At such times the intensity may become several hundred kR, and approach brightness coefficient IV, which is considered the maximum intensity for any aurora. For rays and narrow arcs the intensity is strongly dependent on the direction of observation. When seen from the side, they may be fairly faint, perhaps a few kR, whereas when seen directly along the magnetic field lines they may reach several tens of kR (cf. Romick and Belon 1967). The intensities of the various lines and bands play a major role in the physical interpretation of the optical emission from auroras, and will therefore be discussed comprehensively in the forthcoming chapters.

1.1.3 The Local Auroral Display

The auroral zones are belts around the geomagnetic poles, at roughly 23° from these, where the occurrence of night-time auroras has its maximum frequency for observers on the earth. Auroras also appear regularly inside these zones, but with less frequency and intensity. Equatorwards, the frequency of occurrence falls rapidly as one moves away from the zones. The occurrence of auroras at low latitudes is always associated with enhanced magnetic activity, magnetic storms. On the dayside the aurora occurs nearer to the poles, but it is not so readily observed optically from the ground. At any given instant of time, the aurora occurs along two ovals around the geomagnetic poles. The geomagnetic colatitudes are about 23° on the nightside and 15° on the dayside, but vary with geomagnetic activity. A further discussion of the auroral zones and the auroral ovals is given in Sect. 1.2.2.

In the auroral zone there is a typical development of the auroral display as seen by an observer stationed at a fixed point on the earth. The display usually starts with quiet homogeneous arcs of intensity 10 kR or so, elongated approximately in the geomagnetic $E-W$ direction, with one single arc or several arcs at adjacent latitudes. After some time, possibly hours, the aurora may develop ray structure and take the form of a band, at the same time increasing in intensity. Simultaneously it moves rapidly, with changes in form and intensity, at times increasing to several hundred kR, and the individual structures in the aurora may show apparent speeds in an easterly or westerly direction of several tens of km s^{-1}. In the course of a few minutes the aurora may spread over the entire sky and become weaker and rather irregular. A typical example of such a "break-up" is shown in Fig. 1.7 (see next page), a sequence of auroral pictures taken with an all-sky camera, which photographs the whole sky through a suitable mirror system, usually at regular pre-set intervals and exposure times.

The most spectacular part of the display usually lasts only a few minutes, the whole active period being perhaps half-an-hour or less. Afterwards the sky is more or less covered with a homogeneous surface of weak auroral light. The intensity is usually between 1 and 5 kR,

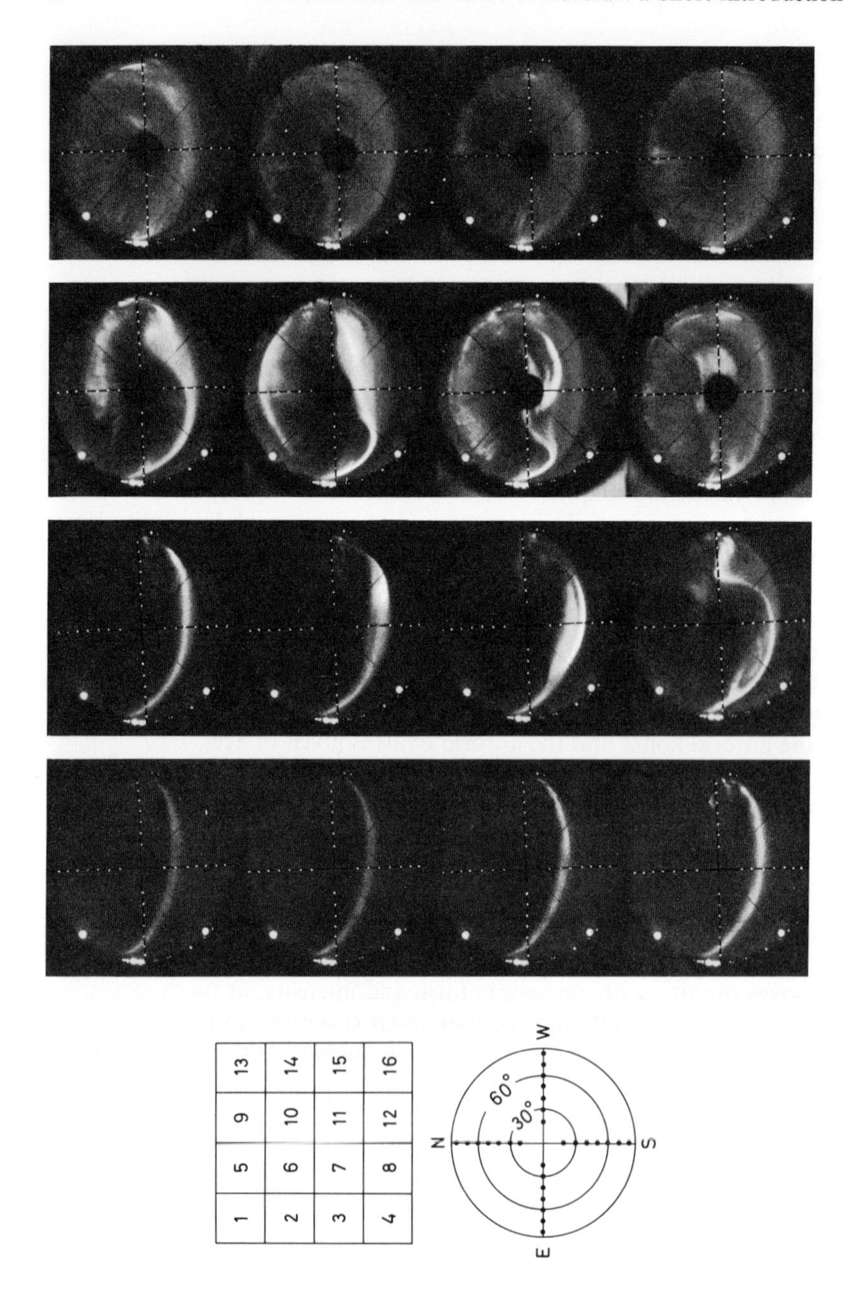

Fig. 1.7 Auroral break-up. Sequence of all-sky camera pictures (Photo: S. Berger). Succession of photographs and orientations are indicated to the left (with zenith distances). 5 s exposure with 40 s intervals

but often subvisual. Weak patches, rays or arcs embedded in this widespread aurora have just sufficient intensity to be visible above the background. Due to contrast effects, the intensity of such weak, widespread surfaces, compared to more discrete and distinct forms, is often underestimated by the eye. This, together with the fact in some cases they really are subvisual, accounts for the fact that such surfaces with background auroral light occur more often than is estimated from visual observations.

If an auroral break-up occurs early in the night, the aurora most often contracts fairly rapidly to quiet arcs again and one or more new active periods may occur during the same night. After geomagnetic midnight*, the auroral pattern is predominantly one of rays and patches, which often show rapid fluctuations in intensity and pulsations.

It must be borne in mind that the instantaneous, global picture of the aurora is governed by the sun and the geomagnetic field, and not by the specific position of the solid earth. Hence, the aurora as seen by an observer at a fixed point on the earth depends both on the development of the overall pattern of the aurora and on the specific movement of the observer underneath this pattern, due to the rotation of the solid earth. We shall return to this question in the next section.

Inside the auroral zone such a fixed observer sees a different pattern. For example, polewards of 75—80° geomagnetic latitude, the arcs seem always to be directed approximately towards the sun (Lassen 1969, Davis 1967). At these latitudes auroras occur more frequently to the sunward side (where it is still dark at midwinter due to the inclination of the earth's axis) than to the nightward side of the geomagnetic axis pole. The violent break-up of quiet forms and vivid displays occurs less frequently poleward of the auroral zones, where short (1—5 min) bursts of forms aligned along the sun-earth line seem to be characteristic (Eather and Akasofu 1969).

The frequency of occurrence as well as the intensity of the auroral displays is clearly correlated with the activity of the sun. This is borne out by the 27-day recurrence tendency correlating with the rotational period of the sun, as well as the obvious 11-year variation correlating with the solar sunspot cycle.

1.2 Auroral Morphology

Only a short account will be given here of the essential features, as the general morphology of auroral displays and their time-history has

---

* For a definition, see Sect. 1.2.1.

recently been described and comprehensively discussed in a book by Akasofu (1968) and also in a review paper by Hultqvist (1969).

The aurora is the result of charged particles interacting with the earth's atmosphere, and the injection of such particles into the atmosphere is due to interaction between the solar wind, i.e. the plasma streaming away from the sun, and the earth's magnetic field. The appearance of auroras in space and time therefore depends mainly on the properties of the solar wind, the geomagnetic field and the density and composition of the atmosphere.

Before we discuss auroral morphology in any detail, it is important to define a suitable coordinate system. To do this, we shall clarify some of the most important properties of charged particles in the earth's magnetic field, and then define a coordinate system suitable for observations and discussion of auroral phenomena.

1.2.1 Magnetic Guiding of Auroral Particles; Coordinate Systems

An electrically charged particle with velocity $\boldsymbol{v}$ moving in a magnetic field $\boldsymbol{B}$ and an electric field $\boldsymbol{E}$ is guided by the basic law of motion

$$m\frac{\mathrm{d}\boldsymbol{v}}{\mathrm{d}t} = e\boldsymbol{E} + e\,\boldsymbol{v}\times\boldsymbol{B} \tag{1.1}$$

in absence of collisions.

The main effect of the $e\boldsymbol{v}\times\boldsymbol{B}$ force is to make a low-energy particle move in a helical trajectory around the field line, if $\boldsymbol{E}=0$ (Fig. 1.8). Størmer (1955) in a monumental classical work made extensive calculations of electron trajectories in the earth's magnetic field. Alfvén (1950) showed that, provided the relative variation of the magnetic induction is small over one revolution of the particle around the field line, the relation

$$\frac{W_\perp}{B} = \mu = \text{constant} \tag{1.2}$$

holds for the (non-relativistic) particle throughout its motion in the magnetic field, $W_\perp$ being the kinetic energy due to the velocity transverse to $\boldsymbol{B}$, and μ the magnetic moment of the spiralling particle. μ is also called the first adiabatic invariant. The particle motion may then be described by a circular path around the so-called guiding center, with the latter moving largely along the field lines. The motion of the guiding center is equivalent to that of a small magnet with magnetic moment μ.

Since magnetic forces cannot alter the total energy of the particle, only changing the direction of the velocity vector, the total kinetic energy W also remains constant if there is no electric field. Therefore, the

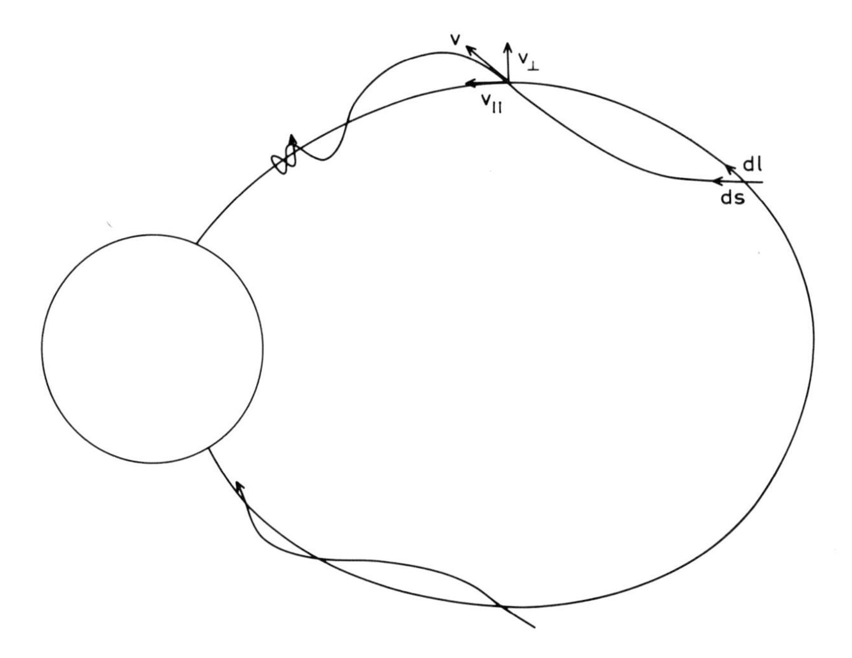

Fig. 1.8 Sketch of particle trajectory in the earth's magnetic field. v, $v_\perp$ and $v_{\parallel}$ are respectively total velocity, velocity component at right angles to, and parallel to, the magnetic field, dl and ds are differentials of path along the magnetic field line and along the particle trajectory, respectively

particle cannot penetrate to regions where the field strength B is larger than B_m, given by

$$B_m = \frac{W}{\mu}. \tag{1.3}$$

When $B = B_m$, $W_\perp$ is equal to W, i.e. the particle moves at right angles to the magnetic field. When the particle has penetrated into the magnetic field to that point, the magnetic forces will make it return to a smaller B, i.e. the particle will mirror at $B = B_m$. Hence, if the magnetic field lines are closed, as in the dipole case, the particle will bounce between two points with $B = B_m$, one in the northern and one in the southern hemisphere.

The second adiabatic invariant is related to the periodic bouncing of a particle between mirror points. It can be shown that the action integral

$$J = 2\int_1^2 m v_{\parallel}\, \mathrm{d}l \tag{1.4}$$

is a constant, provided that the magnetic induction B at any point along the trajectory does not change by any large fraction during the time it takes the particle to move from one mirror point to the other. Here $v_{\parallel}$ is the velocity component and $\mathrm{d}l$ the differential path length component along the field line. The integral is taken from one mirror point to the other. By using the Eqs. (1.2) and (1.3) it is elementary to show that

$$J = 2mv \int_1^2 \sqrt{\left(1 - \frac{B}{B_m}\right)}\, \mathrm{d}l \tag{1.5}$$

or

$$I = \frac{J}{2mv} = \int_1^2 \sqrt{\left(1 - \frac{B}{B_m}\right)}\, \mathrm{d}l\,. \tag{1.6}$$

Hence, $I = J/2mv$ is a characteristic constant for all particles moving along a magnetic field line and mirroring at a particular field strength B_m, provided there are no electric fields or rapid time variations in the magnetic field. I is sometimes called the integral invariant.

In addition to the spiral motion along the field line between the two mirror points, an electrically charged particle also suffers a slow drift around the earth. This drift is caused by the gradient and curvature of the magnetic field; an additional drift may also be caused by electric fields. For a comprehensive treatment of these problems the reader is referred to Vol. 2 of *Physics and Chemistry in Space* (Roederer, 1970) and to other suitable textbooks, for example Hess (1968). It is sufficient to point out here that the first two adiabatic invariants remain constant during the drift motion of the particle, provided this is slow compared to its other motions.

It can be shown that at any longitude there is only one field line which can satisfy the condition that the particle mirrors at a point where $B = B_m$, and that the integral I has a given, constant value. The ensemble of such field lines defines a shell around the earth. This led McIlwain (1966), to define a new parameter L to describe this shell.

Consider a dipole magnetic field: The guiding center follows dipole lines, again preserving B_m and I. These field lines intersect the geomagnetic equatorial plane at a distance R from the center of the earth. The ratio

$$L = R/r_e\,, \tag{1.7}$$

where r_e is the radius of the earth, is a characteristic of the above-mentioned shell, because R is a constant parameter of the path. Hence L is a convenient parameter, in addition to the azimuth (longitude) and B_m, to characterize the path of the particle in a dipole field. The shell is called an L-shell. Mathematically, in a dipole field, L can be accurately related to I and B_m, i.e.

$$L = L(I, B_m)\,. \tag{1.8}$$

For a non-dipole field the true value of R for the particle will vary with azimuth. One still can define L by the same function (1.8) of I and B_m for the particular field line in question and use it as a field line parameter. In that case, however, it does not satisfy (1.7).

Since I is determined by B_m and B only, particles with the same mirror point will describe the same L-shell if their paths coincide at one longitude. Particles with different B_m, however, do not necessarily describe the same shell in an asymmetric field. We consider two particles initially on the same field line, but with different mirror points. When drifting past a given longitude, they may not necessarily be on the same field line, because both have to satisfy their own requirements regarding I as well as B_m.

This effect is called L-shell splitting. For $L \leqslant 4$ it is very small, about 1 per cent. At $L = 5$ it may amount to about 10 per cent, and for larger values of L it is considerable (Roederer 1968). This is due to the increasing asymmetry between the noon and midnight sides of the magnetosphere with increasing distance from the earth. The effect of shell splitting has been quantitatively tested by satellite observations (Pfitzer *et al.* 1969).

In a dipole field, a field line characterized by L crosses the earth's surface at a latitude Λ_L given by

$$\cos^2 \Lambda_L = \frac{1}{L}\,. \tag{1.9}$$

Λ_L is called the invariant geomagnetic latitude, and is a quantity which may be used instead of L to characterize a particle shell. Λ_L is constant along the intersection line between the L-shell and the earth's surface.

As borne out by Eq. (1.8) and the discussion of L-splitting, the definition of L for a particular field line depends on the mirror point B_m of the particle considered. For auroral particles it is convenient to define L by particles mirroring in the earth's atmosphere, at auroral heights.

Although it turns out that the parameter L or Λ_L is convenient for describing trapped particles, its physical significance relative to particles which penetrate into the atmosphere, and hence to visual auroras, is not entirely clear. The injection of auroral particles into the atmosphere is the result of a complicated interaction between the solar wind and the earth's magnetic field. This field is strongly deformed in the outer region, with a tail on the side away from the sun, and a neutral sheet in this tail. Particles enter the geomagnetic field through the tail and begin to drift around the earth under influence of an electric field set up by the interaction between the solar wind and the geomagnetic field. Subsequent betatron acceleration and release of quasitrapped particles by instabilities may account for the auroral particles. It is beyond the scope of this book to discuss this process in detail, but we shall note that the outer magnetic field is strongly disturbed by this interaction and by the ring current set up by the drifting particles. This makes the significance of L or Λ_L dubious at high latitudes. For L larger than 7 or 8, the magnetic field varies so much during geomagnetic storms that the values computed for quiet fields may be misleading. This effect has been demonstrated in studies of auroras conducted at normally conjugated points in the north and in the south (i.e. points at the same magnetic field line under quiet conditions) which show how the conjugacy of auroras breaks down (Nielsen *et al.* 1970). Still, the value of Λ_L is the best coordinate we have to describe the magnetic latitude.

Kilfoyle and Jacka (1968) have pointed out that the true shape of the magnetic field lines also makes it necessary to apply similar corrections to geomagnetic longitude, since a particular magnetic field line does not lie in a plane, as happens in the dipole approximation.

Since the storm phenomenon is controlled by the sun and the geomagnetic field, the direction to the sun and the direction of the dipole axis of the earth are the most relevant directions. A suitable coordinate system is therefore the following: The principal axis is the dipole axis of the earth's magnetic field. The plane through this axis and the earth-sun line is a principal reference plane. Local magnetic time at a particular point is defined by the angle between this plane and that through the dipole axis and the point in question. These two planes coincide where the magnetic time is 1200 h (magnetic midday) or 0 h (magnetic midnight). The third coordinate is either L, Λ_L or the dipole latitude Φ. The preference here depends on the phenomenon under study. Raw data from auroral observations are often presented using the dipole latitude, since proper conversion to another coordinate depends on detailed knowledge about the field.

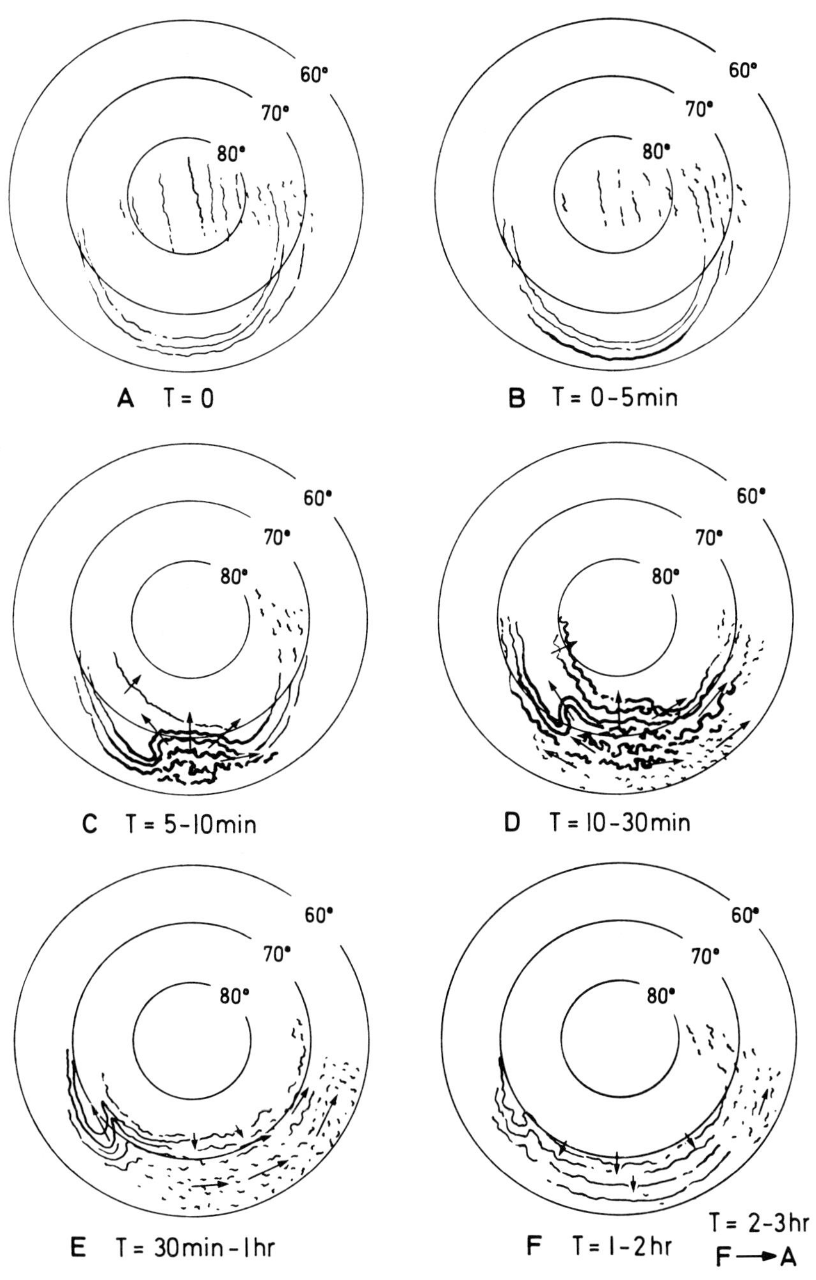

Fig. 1.9 The development of an auroral substorm. North geomagnetic pole in the center and the sun towards top of diagram (Akasofu 1964, courtesy Pergamon Press)

1.2.2 Worldwide Auroral Morphology

Much of our systematic knowledge about the auroral display, organized in a dipole-local magnetic time coordinate system, is due to Davis (1962, 1967) and Akasofu (1965, 1968).

A single aurorally active period is often called an auroral substorm. An illustration of the development of an auroral substorm in the dipole local time coordinate system is given in Fig. 1.9. Here each diagram has the geomagnetic axis pole as origin, the axis being vertical to the plane of the paper, and the direction to the sun at the top. To begin with, a reasonably quiet system of auroral arcs or bands exists, such as shown in plot A. The activity starts with a brightening of one or more of the bands (B), followed by a deformation and poleward movement of the entire arc system around the midnight meridian. New rayed forms are created in place of the original arcs (C) and the maximum phase (D) is characterized by bright forms in rapid movement and with rapid intensity variations all along the auroral oval. Phases (A) to (D) correspond to the sequence shown in Fig. 1.7. The original bands and the newly created forms generally move away from the initial disturbance center. After the time of maximum of the substorm, weaker and more diffuse forms occur (E and F), and the aurora gradually weakens and again takes on the shape of arcs and bands.

It is easily seen that the apparent development for an observer at a given point on earth will depend on his position (each observer sees a circular area approximately 1000 km in diameter at 100 km altitude). Altogether, his impression will depend on both the development of substorms in the course of the night and his own travel underneath the auroral pattern.

In the evening sector, arcs and bands will dominate, whereas in the morning hours diffuse patches, single rays and other scattered, weak auroral phenomena are most common. The magnitude and intensity of the substorm may vary greatly, some substorms occurring as mere bulges in the arc system, while others may disrupt it for several hours. The usual time interval between two substorms ranges from 3 to 12 hours.

Each substorm is associated with disturbance current systems in the ionosphere, giving rise to characteristic magnetic disturbances (Akasofu 1968).

When studying the instantaneous occurrence of aurora in a dipole-local time coordinate system, one obtains an oval-shaped belt of maximum auroral activity which is asymmetric around the geomagnetic pole; this is called the auroral oval. During moderate activity it is located about 23° and 15° from the geomagnetic pole on the night- and day-sides, respectively, and it widens with increasing geomagnetic activity (cf. Akasofu 1968 for a comprehensive discussion of auroral morphology and substorms). The orientation of single auroral arcs seems to follow this auroral oval, too (cf. Gustafsson *et al.* 1969, Sievwright 1969). Akasofu (private communication) recently arrived at a similar conclusion through observations from an airplane flying along the oval. Moreover, auroras on the dayside of the oval more frequently show red color, indicating greater heights (Akasofu 1970). It seems established that there is a zone of weak, soft-particle precipitation in and near the oval at the dayside, stretching to about 5° equatorwards from the oval, and another zone of harder particles equatorwards of this (cf. Harz and Brice 1967, Romick 1970, Whalen 1970).

A general conjugacy of occurrence of auroral phenomena in the northern and southern hemispheres seems to exist (cf. Akasofu 1968, Bond 1968, Davis *et al.* 1969, Nielsen and Davis 1970, Nielsen *et al.* 1970), although the details imply large disturbances in the geomagnetic field and distortions of the quiet-time position of the conjugate point.

The auroral zones, which earlier were studied with much enthusiasm, are the zones of maximum frequency of auroras as seen by observers at a fixed point on the earth. However, this type of statistics in fact integrates the aurora measured at any geographic point over both local magnetic time and universal time. If the earth's magnetic field were a dipole one, it should leave a variable depending on geomagnetic latitude only (Omholt 1965). The fact that the classical auroral zones are not circular reflects that the dipole approximation is not entirely valid. For this reason L or Λ_L are better coordinates to use than the dipole latitude. In fact, the auroral zones follow $L=$constant lines to a very good approximation.

Another complication in this simple way of describing and studying auroral substorms arises from the fact the sun usually is not in the geomagnetic equatorial plane. Due to the inclination of the earth's axis of rotation with respect to the earth's orbital plane (23°) and to the angle between the dipole axis and the axis of rotation (11°), the direction to the sun may deviate as much as $\pm 34°$ from the geomagnetic equatorial plane during a year. These two effects give rise to diurnal and seasonal variations which can be studied statistically if a proper separation of data is accomplished.

The most important variation is probably the seasonal one, due to the angle between the geomagnetic dipole axis and the earth-sun direction. Variation of the oval with geomagnetic axis direction has been observed by Feldstein and Starkov (1970).

For discussions of single auroral events, it may also be useful to note that the variation with universal time T also includes components originating in the 11-year solar cycle and a 27-day recurrence tendency due to the period of revolution of the sun.

1.2.3 Mid-Latitude Red Arcs

There are some particular high red arcs, also called M-arcs, which occur at middle latitudes and which probably must be classified as auroras. A red arc consists of a subvisual, broad arc elongated in the geomagnetic east-west direction, and it is monochromatic (above background) in the red [OI] lines at 6300 and 6364 Å. These arcs were first detected by Barbier (1958) in 1958, and studied intensively in the following years, in particular by F. E. Roach and associates (cf. Roach and Roach 1963, Elvey 1965, Marovich 1966, Roach and Smith 1967, Hoch *et al.* 1968, Glass *et al.* 1970, Roble *et al.* 1970, Hoch and Clark 1970). Their height is well above what is common for polar auroras, the lower broder being around 300 km and the upper border in the vicinity of 700 km. The latitudinal extent may be as much as 600 km. The intensities of the 6300 Å emission in these arcs are usually of the order 1—10 kR, which at this wavelength is below the threshold of visibility.

The following properties are characteristic of these arcs (Roach and Smith 1967):

1. They have been observed between geomagnetic latitudes 41° and 60° with a median position of 53°.

2. They are about 600 km in north-south extent.

3. The spread in height is from about 300 to 700 km with the photometric center near 400 km.

4. They extend east-west for thousands of kilometers, possibly right round the earth.

5. There is some evidence that they occur simultaneously in the northern and southern hemispheres.

6. They are oriented along magnetic, not geographic, parallels. The orientation is better described by the magnetic invariant latitude than by either geomagnetic or dip latitude.*

7. Only rarely have they been intense enough to be visible. The visibility threshold at 6300 Å is about 10 kR, and a typical integrated intensity is 6 kR.

8. The intensity is positively correlated with geomagnetic activity.

---

* Exceptions have been described (Glass *et al.* 1970).

9. Present observational evidence, though covering barely more than one complete solar cycle, gives a positive correlation between their occurrence and sunspot activity.

10. The north-south movements are slow, but often measurable.

11. The life of a typical arc is about one day—only rarely has an arc been observed on successive days.

12. An HF radio wave from a satellite to a ground station on traversing an arc has been observed to suffer extensive scintillation. Scintillation has also been noted for radio-frequency beams from radio sources which traverse an M-arc.

13. On one occasion, an arc existed directly beneath regions of enhanced outer zone Van Allen radiation.

Ichikawa and Kim (1969) and Ichikawa, Old and Kim (1969) found evidence that the M-arcs are associated with visible, distinct auroras 5—10° poleward of the M-arcs. Both forms were moving, and the average N—S speed of the distinct auroras was about 3.5 times that of the M-arc. Moreover, directions of movement correlated, with the M-arc changing its direction 1—3 min after the visible aurora.

1.2.4 Polar Glow and Mantle Auroras

The polar glow and mantle auroras are particular types of glow, extending over large areas of the polar cap (cf. Sandford 1967a, b). The polar glow auroras are excited by solar cosmic rays, in the energy range 1—100 MeV, and are associated with strong radio absorption events, the so-called PCA's (polar cap absorption, cf. Sect. 8.6). A uniform glow is generated over the whole polar cap down to a geomagnetic latitude of about 60°, where the cosmic ray cut-off limits penetration. The intensity rises steadily for about a day, the maximum intensity of the N_2^+ first negative band being up to 10 kR. The intensity of the glow then decays slowly over a period of a few days. The onset of the magnetic storms associated with these events occurs near the time of maximum intensity.

The penetration altitudes of the particles range from 20 to 100 km, at which heights the glow occurs. A further discussion of these auroras is given in Sect. 8.6.

There is another type of glow in the auroral region, detected by Sandford, who called it mantle aurora. This is an extensive auroral glow with a distribution differing considerably from that of a discrete aurora (cf. Sandford 1967, a, b). The maximum intensity occurs in the morning hours, and varies with the level of geomagnetic activity, whereas the size of the region apparently does not change. The diurnal maximum intensity is usually between 1 and 10 kR, while the minimum intensity is about 5 times less. Sandford's measurements were made during the

International Geophysical Year, with high solar activity. Eather (1969), operating from a high-altitude aircraft in 1968, failed to find such an emission at 80° N invariant latitude. From observations in Antarctica in 1963 Sandford (1968) also found a lesser intensity, and this strongly suggests that the intensity of the mantle aurora, and perhaps its extension, depend on solar activity.

1.3 Particle Behaviour: Reflection, Absorption and Scattering

1.3.1 Mirror Effects

We shall not go into a detailed discussion of the injection mechanisms for auroral particles, but restrict ourselves to some properties and aspects of these mechanisms which are important to render optical observations useful. The magnetic mirroring of particles is in this respect of limited importance but, in order to study the properties of the precipitating particles from optical data, it is desirable to have some basic knowledge about particle trajectories in, and just above, the atmosphere.

An important basic theorem is that of Liouville, which states that the density of particles in phase space is constant along the trajectory of a particle. This implies that if, at a given field line, one has an isotropic angular distribution at some distance from the earth, the distribution remains isotropic as the particles approach the earth, and the flux of particles is independent of the convergence of the magnetic field. (The decrease in flux tube area is compensated by the reflection or mirroring of particles with large pitch angle.) Therefore, the relation between the auroral luminosity and the particle flux further out in space is independent of local anomalies in the magnetic field. Further, the requirement of isotropy comprises only those particles which reach the atmosphere, and not those which are reflected higher up. However, another matter is that the converging of the magnetic field affects the relation between the size of a luminous auroral area and the size of the corresponding beam cross section in space. Hence, independently of the angular distribution, the luminosity integrated over the aurora may in any case be sensitive to the local B-value and mirror properties.

If there is strong anisotropy, however, for particles which are not reflected before they reach the atmosphere, the local magnetic anomalies could affect the ratio between the local auroral intensity and the corresponding particle density in space.

The mirroring of particles within the atmosphere usually is not very important except when particles with large pitch angles are involved. The main auroral luminosity is generally confined to a height interval of 100 km or less, over which B varies less than 5%. This should give an upper limit for the fraction of particles that mirror in the atmosphere. The mirroring of particles therefore causes second-order effects only, and is probably of importance only in specific cases. Two such cases may be the detailed and accurate study of the top part of long auroral rays, and of the small red-shifted part of the zenith Doppler profile of the hydrogen Balmer lines.

1.3.2 Absorption and Scattering of Particles in the Atmosphere

When fast electrons and protons penetrate into the atmosphere, they suffer collisions, elastic as well as inelastic. The inelastic collisions, and the secondary processes which follow, cause excitation and ionization of atmospheric atoms and molecules. These processes will be described in detail in the following chapters.

The primary particles are gradually slowed down until they are brought to rest by collision processes which also cause changes in the pitch angle of the particles. A detailed study of the relevant processes is necessary to relate the initial properties such as energy and pitch angle distributions of a beam of primary particles before it enters the atmosphere to the resulting auroral luminosity and enhanced ionization in the ionosphere.

For electrons, both scattering and energy loss due to inelastic collisions need to be considered to obtain a reasonably accurate picture, The much heavier protons, however, suffer only small angular scattering during collisions, and the resulting change in pitch angle will be small. Since particles will continue to spiral around the same magnetic field line regardless of changes in pitch angle, the geometry of the aurora will reflect the cross-sectional geometry of the beams, with the limitations given by the size of the gyroradii, which for auroral primaries typically are of the order of 1 to 10 m for electrons and 100 to 1000 m for protons.

With sufficiently detailed knowledge of the basic physical processes involved, the atmosphere may be used as an analyzer of the beam of primary particles. Uncertainties will appear, however, due to the fact that both pitch angle and energy determine at what height in the atmosphere the energy of the particle is deposited. Nevertheless, useful information may be gained from studies of the luminosity of auroras, its height distribution and geometry.

References

Akasofu, S.-I.: J. Geophys. Res. **68**, 1667 (1963).
— Planetary Space Sci. **12**, 273 (1964).
— Space Sci. Rev. **4**, 498 (1965).
— Polar and magnetic substorms. Reidel Publ. Co. 1968.
— Trans. Am. Geophys. Union **51**, 370 (1970).

Alfvén, H.: Cosmical Electrodynamics. Oxford: Clarendon Press 1950.
Barbier, D.: Ann. Geophys. **14**, 334 (1958).
Bond, F.R.: Australian J. Phys. **22**, 421 (1968).
Chamberlain, J.W.: Physics of the Aurora and Airglow. Academic Press 1961.
Chapman, S.: In Aurora and Airglow. (Ed. B. M. McCormac) Reinhold Publ. Co. 1967.
— In Atmospheric Emissions. (Ed. B. M. McCormac and A. Omholt) Van Nostrand Reinhold Co. 1969.
Davis, T.N.: J. Geophys. Res. **67**, 59 (1962).
— — In Aurora and Airglow (Ed. B. M. McCormac) Reinhold Publ. Co. 1967.
— Nielsen, H.C.S., Olson, H.C., Langlotz, R.J.: In Annual Report. Geophysical Institute, Alaska **1968–69** (1969).
Eather, R.H.: J. Geophys. Res. **74**, 153 (1969).
— Akasofu, S.-I.: J. Geophys. Res. **74**, 4794 (1969).
Elvey, C.T.: In Auroral Phenomena. (Ed. M. Walt) Stanford 1965.
Feldstein, Y.I., Starkov, G.V.: Planetary Space Sci. **18**, 501 (1970).
— Khorosheva, O.V., Lebedinsky, A.L.: J. Physical Soc. Japan **17**, Suppl. A. 249 (1962).
Glass, N.W., Wolcott, J.H., Wakefield, R.L., Peterson, R.W.: Airborne observations of the night airglow. Preprint LA-DC-10557. Los Alamos Scientific Lab. 1970.
Gustafsson, G., Feldstein, Y.I., Shevnina, N.F.: Planetary Space Sci. **17**, 1657 (1969).
Hartz, T.R., Brice, N.M.: Planetary Space Sci. **15**, 301 (1967).
Hess, W.N.: The Radiation Belt and Magnetosphere. Blaisdell Publ. Co. 1968.
Hoch, R.J., Clark, K.C.: J. Geophys. Res. **75**, 2511 (1970).
— Marovich, E., Clark, K.C.: J. Geophys. Res. **73**, 4213 (1968).
Hultqvist, B.: Rev. Geophys. **7**, 129 (1969).
Ichikawa, T., Kim, J.S.: J. Atmospheric Terrest. Phys. **31**, 547 (1969).
— Old, T., Kim, J.S.: J. Geophys. Res. **74**, 5819 (1969).
International Auroral Atlas: Edinburgh University Press 1963.
Kilfoyle, B.P., Jacka, F.: Nature **220**, 773 (1968).
Kim, J.S., Currie, B.W.: Can. J. Phys. **36**, 160 (1958).
Lassen, K.: In Atmospheric Emissions. (Ed. B. M. McCormac and A. Omholt) Van Nostrand Reinhold Co. 1969.
Marovich, E.: ESSA Tech. Rep. IER 16–ITSA 16 Boulder 1966.
McIlwain, C.E.: Space Sci. Rev. **5**, 585 (1966).
Nielsen, H.C.S., Davis, T.N.: Trans. Am. Geophys. Union **51**, 395 (1970).
— — Glass, N.B.: Report to the 51st Annual Meeting of AGU, April 1970 (Conjugacy in the gross structure of visual aurora during magnetically disturbed periods) 1970.
Omholt, A.: In Introduction to Solar Terrestrial Relations. (Ed. J. Orther and H. Maseland) Reidel Publ. Co. 1965.
Pfitzer, K.A., Lezniak, T.W., Wincker, J.R.: J. Geophys. Res. **74**, 4687 (1969).
Roach, F.E., Roach, J.R.: Planetary Space Sci. **11**, 523 (1963).
— Smith, L.L.: In Aurora and Airglow. (Ed. B. M. McCormac) Reinhold Publ. Co. 1967.
Roble, R.G., Hays, P.B., Nagy, A.F.: Planetary Space Sci. **18**, 431 (1970).
Roederer, J.G.: In "Earth's Particles and Fields". (Ed. B. M. McCormac) Reinhold Publ. Co., p. 193 1968.
— Dynamics of Geomagnetically Trapped Radiation. Berlin-Heidelberg-New York: Springer 1970.

Romick, G.J.: Trans. Am. Geophys. Union **51**, 370 (1970).
— Belon, A.E.: Planetary Space Sci. **15**, 475 (1967).
Sandford, B.P.: In "Aurora and Airglow". (Ed. B. M. McCormac) Reinhold Publ. Co. 1967a.
— Space Res. **7**, 836 (1967b).
— J. Atmospheric Terrest. Phys. **30**, 1921 (1968).
Sievwright, W.M.: Planetary Space Sci. **17**, 421 (1969).
Störmer, C.: The Polar Aurora. Oxford: Clarendon Press 1955.
Whalen, J.A.: Trans. Am. Geophys. Union **51**, 370 (1970).

Chapter 2

The Electron Aurora: Main Characteristics and Luminosity

2.1 Introduction

In this section we shall discuss the use of auroral luminosity data as a means of studying the primary electrons producing the aurora. That is, we shall study the properties of the atmosphere as a detector for auroral electrons.

It is well known that protons contribute to the production of auroral light, but this problem is deferred to Chapt. 3. Here we shall restrict ourselves to a consideration of electrons only. We shall first discuss the light production by energetic electrons in air, from a theoretical as well as an experimental point of view, and then use this information to deduce data on the primary electrons from auroral luminosity and height distribution in the atmosphere*. Finally, a description of the geometry and dynamics of the electron aurora will be given.

The auroral spectrum contains a number of emission lines and bands which arise mainly from atomic and molecular nitrogen and oxygen. A description of the spectrum and a discussion of the main excitation mechanisms are given in Chapts. 4 and 5. However, two important features should be emphasized here. One is that the first negative N_2^+ bands (among which those at 3914 and 4278 Å are the strongest) most certainly are directly related to ionization by electrons in air. This emission, induced by an electron beam in air, is known in great detail. The other important feature is that the height distribution of the strong [OI] line at 5577 Å is very similar to that of the first negative bands, though perhaps with some exceptions.

---

* An artificial electron aurora has been generated with a rocket-borne accelerator (cf. a series of abstracts in Trans. Am. Geophys. Union 1970, Vol. 51, p. 394 and 395). This may eventually provide a useful tool for a more direct study of the effects of fast electrons in the upper atmosphere.

2.2 N_2^+ Emission and Ionization

It is well established that fast electrons in air produce about three ion pairs for each hundred electron volts of initial energy (Dalgarno 1962). Thus, a 1 keV electron produces about 30 ion pairs before it is brought to rest. Ionization of nitrogen

$$N_2 + e \rightarrow N_2^+ + e + e \tag{2.1}$$

leaves some of the nitrogen molecules in the excited state $B^2\Sigma$, the upper state of the first negative bands. Provided it does, this results in emission of a photon in the strong λ 3914 or λ 4278 bands in a certain fraction of cases. The relevant cross-sections, *i.e.* that for ionization into any state of excitation of the ion and that for ionization followed by emission of a λ 3914 or a λ 4278 photon, have been extensively studied in the laboratory (Stewart 1956, Hayakawa and Nishimura 1964, Hartman and Hoerlin 1962, McConkey and Latimer 1965, Sheridan, Oldenberg and Carleton 1961, McConkey *et al.* 1967, Aarts *et al.* 1968, Srivastava and Mirza 1968, Hartman 1968, Khare 1969, cf. also discussions by Dalgarno *et al.* 1965 and Davidson 1966).

The most probable value for the ratio between the total rate of ionization in the atmosphere, including ionization of oxygen, and the emission of λ 3914 photons is about 25. For the λ 4278 and λ 4709 bands the corresponding ratios will be about 75 and 300, adopting a ratio of 1.0:0.34:0.075 for the emissions in the λ 3914, λ 4278 and λ 4709 bands respectively (cf. Srivastava and Mirza 1968). These ratios vary very little with the energy of the primary particles because the ratio between the cross-sections in question varies only slightly with electron energy. For an aurora excited mainly by protons as primary particles, the situation is different, cf. Sect. 3.2.5.

The ratio between the total influx of energy by fast electrons and the resulting emission of λ 3914 photons is then approximately 8×10^2 eV per photon. Using the λ 4278 band, the corresponding ratio is slightly higher than 2×10^3 eV per photon. This efficiency is in good agreement with direct measurements in the aurora, which gave about $(1.1 \pm 0.4) \times 10^{-3}$ for converting primary energy to λ 4278 light (Bryant *et al.*, 1970).

Hence it is possible to determine the total energy flux carried into the atmosphere by the electrons by measuring the emission of λ 3914 or λ 4278 photons from the aurora. Certain practical problems arise, however, because any optical measurement from the ground in effect involves integration along the line of sight. A comparison is therefore strictly valid only when the optical observations are made in the direction along the magnetic field lines so that one integrates all the light produced by the ensemble of particles following the field lines within the optical field of view. Nevertheless, giving due consideration to the overall geometrical pattern of the aurora, useful measurements over a fairly large area may be obtained from one observation point.

Rocket-borne instruments may provide information on the height distribution of luminosity, and hence that of the rate of ionization. Simultaneous measurements of the ionospheric electron density may yield effective recombination coefficients.

Using rocket measurements, Ulwick (1967) compared the electron production computed from (a) the primary electron flux in auroras and (b) the intensity of the λ 3914 band, measured simultaneously. The agreement between the two sets of data was excellent for a bright aurora, but less good for a weak one. Comparison of the electron production rate computed from the λ 3914 intensities and the ionospheric electron densities measured simultaneously gave recombination coefficients α from 4 to 7×10^{-7} cm^3 s^{-1} at heights between 114 and 125 km. The consistency of the values derived for α was excellent, considering that the weakest and strongest aurora differed by as much as a factor of 40 in intensity. Ulwick also found that the relation between the flux of primary particles and the ionospheric electron densities measured by rockets, and the λ 5577 brightness from ground measurements, was reasonably consistent with the theoretical results obtained by Dalgarno *et al.* (1965).

Similarly, McDiarmid and Budzinski (1964) compared values for the ionization rate and the electron density as a function of height from 82 to 95 km, obtaining a recombination coefficient decreasing from 10^{-5} cm^3 s^{-1} at 82 km to about 10^{-7} cm^3 s^{-1} at 95 km (cf. also Anger 1967).

In recent rocket measurements from Andøya in Norway (67° geom. lat.), the height distributions of the λ 4278 intensity and the local electron density were measured simultaneously (Jespersen *et al.* 1969, Bryant *et al.* 1970). The effective recombination coefficient α was slightly less than 10^{-7} cm^3 s^{-1} above 115 km, increasing to slightly above 2×10^{-7} cm^3 s^{-1} at 100 km. There is reasonable agreement between results from 3 flights, between which the intensity of the aurora differed by as much as a factor of 20. Similar measurements by Baker (1968) gave values of 3 to 6×10^{-7} cm^3 s^{-1}, whereas Matthews and Clark (1968) found about 10^{-7} cm^3 s^{-1}. Data obtained by McNamara (1969) are consistent with a value of about 2×10^{-7} cm^3 s^{-1}.

The value of a few times 10^{-7} cm^3 s^{-1} is not contradictory to estimates obtained from comparison of electron densities in the E-layer during aurora (about 100 km altitude) with luminosity, measured respectively by radio echo techniques and optical observations from the ground (cf. Sect. 8.3). The measured values for α are of the same order of magnitude as laboratory values for NO^+ and O_2^+, the two most probable positive ions in the height region of interest (cf. Biondi 1969). It has become apparent, however, that the ion chemistry in the E-layer may be rather complicated (Swider and Narcisi 1970).

In recent years, the quantitative relation between flux of particles and output of light has also been studied by satellite techniques, although not yet with sufficient accuracy to improve the quantitative relations deduced from laboratory and theoretical data. A recent review of satellite measurements has been given by O'Brien (1967).

2.3 Theoretical Height Distribution of the First Negative N_2^+ Bands

If the energies and pitch angles of the incident electrons are known, the height distribution of the auroral luminosity can be computed, provided the relevant cross-sections are known with sufficient accuracy. Similarly, properties of the energy and pitch angle distribution may be studied from the height distribution of auroral light. This comparison is not straightforward, however, since there are two variables in the theory and only one in the observations. As stated in the preceding section, the first negative N_2^+ bands are directly related to the ionization. Therefore, the emission of these bands can be estimated from the known properties of the electrons. For other emissions the relation between the primary particles and the emission is more uncertain (cf. Chapt. 5).

Computing the ionization performed by an electron along its path in air, one finds that highest efficiency occurs at a very short distance from the terminal of the path, at a point where the electron has an energy of about 150 eV. In accordance with what was stated in the preceding section, the luminosity along the path should be distributed in similar fashion. This, together with the exponential increase of atmospheric density, gives a very sharp and concentrated luminosity distribution with height for monoenergetic electrons impinging upon the atmosphere along magnetic field lines.

The picture is considerably complicated, however, by the fact that electrons move in helical paths in the presence of a magnetic field. Inelastic as well as elastic collisions cause discontinuous changes in the direction of the velocity vector of the electron. Since each electron suffers a great number of these collisions during its lifetime, the scattering effect is important, in fact, a dominating factor in shaping the luminosity curve as a function of height in the atmosphere.

This problem has been studied theoretically by Spencer (1959) and experimentally by Grün (1957). Chamberlain (1961) has given an excellent review of Spencer's theory, applied to particles penetrating into the atmosphere. We shall therefore confine ourselves to the physical aspects and their application to the aurora, such as that carried out by Rees (1963, 1964a, b, 1969), and by Belon *et al.* (1966).

When a parallel beam of electrons is launched into a vessel of nitrogen or air at constant pressure, the individual electrons are scattered out from the original beam direction and a diffuse glow of light is produced. Grün (1957) made such experiments in nitrogen, with electron energies varying from 5 to 54 keV, and measured the resulting emission of first negative bands. If a magnetic field were introduced, as is the case in the earth's atmosphere, the electrons would not be significantly scattered away from the field line along which they enter, because they are forced to spiral around a field line between each collision. Hence the luminosity is confined to a narrow tube parallel to the magnetic field*. But since the sole effect of the magnetic field is to wind the path in a spiral around the field lines, the luminosity distribution along such a tube is the same as it would be if the luminosity, as distributed in the field free case, were projected on to the magnetic tube. Since this tube is narrow in width compared to the length over which the electrons deploy their energy, it can for all practical purposes be considered as a single field line, with the integration performed over a layer at right angles to the direction of the magnetic field.

Let R be the average range of an electron in absence of scattering, i.e., the penetration depth measured along the individual path (measured in g cm^{-2} along the path). Since for an individual electron the actual penetration depth is the result of statistical processes, there will be a small statistical variation of the penetration depths, a straggling of particles with the same original energy. If z is the depth projected along the magnetic field, the property of interest in our case is the luminosity integrated over a layer normal to this direction as a function of z. According to Spencer (1959) and Grün (1957), this luminosity as a function of z/R happens to be very similar for all electron energies within the range studied (5—54 keV). An average curve, adopted by Rees (1963), is shown in Fig. 2.1. The curve labelled "monodirectional" represents particles penetrating into the atmosphere along the magnetic field line. The part towards negative z/R is the hypothetical curve from

* The gyroradius of electrons of a few keV energy in the earth's atmosphere is a few meters or less. Since the average scattering angle is small, the electron will, after a collision, spiral around a field line which on the average is less than one meter away from that around which it spiralled before the collision. This effect will tend to diffuse a narrow beam of electrons arriving along one particular field line, but only to a few tens of meters in diameter. This is confirmed by the narrowness of auroral rays (cf. Sect. 2.6).

back-scattered electrons, and represents the fraction of energy which is scattered out into space again. As expected, this fraction is relatively larger for a wider angular distribution, since electrons arriving with large pitch angle θ need only small scattering angles to make them move upwards again. Stadsnes and Maehlum (1965) and Walt (1967) have considered back-scattered electrons explicitly. To illustrate the importance of scattering, we quote a few figures from Stadsnes and Maehlum's detailed computations. The fractions of back-scattered electrons, when the electron energy is between 20 and 100 keV, are about 20%, 40% and 70% for $\theta = 30°$, $60°$ and $80°$, respectively.

Assuming an isotropic pitch angle distribution for the electrons between $\theta = 0$ and $\theta = \theta_m$, Rees (1963) computed the normalized energy dissipation $\lambda(z/R)$ for $\theta_m = 70°$ and $80°$ as well as for an angular distribution varying as $\cos\theta$. These results are also shown in Fig. 2.1.

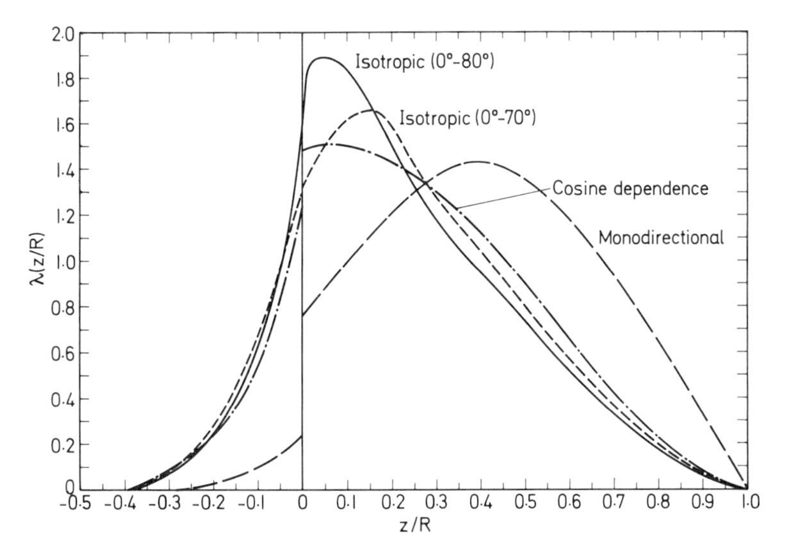

Fig. 2.1 Normalized energy dissipation distribution function for four angular dispersions of the incident electron stream (Rees 1963, courtesy Pergamon Press)

If the density variation of the atmosphere with height is known with sufficient accuracy, the luminosity distribution with height of the first negative N_2^+ bands may now easily be derived. The fact that the atmospheric density gradient is not parallel to the magnetic field does not disturb the picture significantly, since the electrons arriving along a particular field line are confined to a flux tube of a few tens of meters in diameter. The air density differs negligibly over such distances. Hence the emission per unit residual range as derived by Spencer and measured by Grün for the entire beam, including scattering, can easily be converted to luminosity per unit length along the field line, or per unit height in the atmosphere, by using the atmospheric density variation with height.

The luminosity distribution as a function of range along the magnetic field is described by:

$$\lambda = \lambda(z/R). \tag{2.2}$$

z would be a geometric distance if the air density were constant along the path. In the atmosphere, however, the emission per unit path l along the field line is

$$L(l) = \lambda(z/R)\rho/R, \tag{2.3}$$

where $z = \int \rho \, dl$, ρ being the atmospheric density. The emission per unit height h in the atmosphere is

$$L(h) = L(l)\frac{dl}{dh} \tag{2.4}$$

where $\frac{dl}{dh}$ is unity for a vertical magnetic field.

$\lambda(z/R)$ is shown in Fig. 2.1 for different angular distributions of the electrons. Converting to the atmosphere, Rees (1963) computed the ionization rate as a function of height, shown in Fig. 2.2, for electron energies between 0.4 and 3000 keV. The numbers in Fig. 2.2 for the ionization rate should be divided by 25 to obtain the approximate λ 3914 emission (Rees used an older figure of 50 for this ratio, which affects the absolute values only). The same $\lambda(z/R)$ curve was used in the entire energy range, although Grün's (1957) measurements covered the range 5 to 54 keV only. Little error is probably introduced by the extension; at the upper limit of 300 keV radiative losses are still unimportant compared to ionization losses, while at the lowest energy the loss per inelastic collision is still small compared to the total energy. To obtain a very accurate luminosity distribution one has to allow for the variation in relative content of N_2 in the atmosphere with height (cf. Rees 1963, Fig. 4).

In addition, Rees (1963) computed emission height profiles from electrons with exponential energy distributions and isotropic angular distribution. He considered the magnetic field lines in the auroral zone to be vertical, rather than having an inclination of about 80°. In a subsequent paper (Rees 1964a) he showed that the correction needed was small.

For a further discussion of auroral electrons, the reader is also referred to a review paper by Rees (1969). Recently, detailed computations

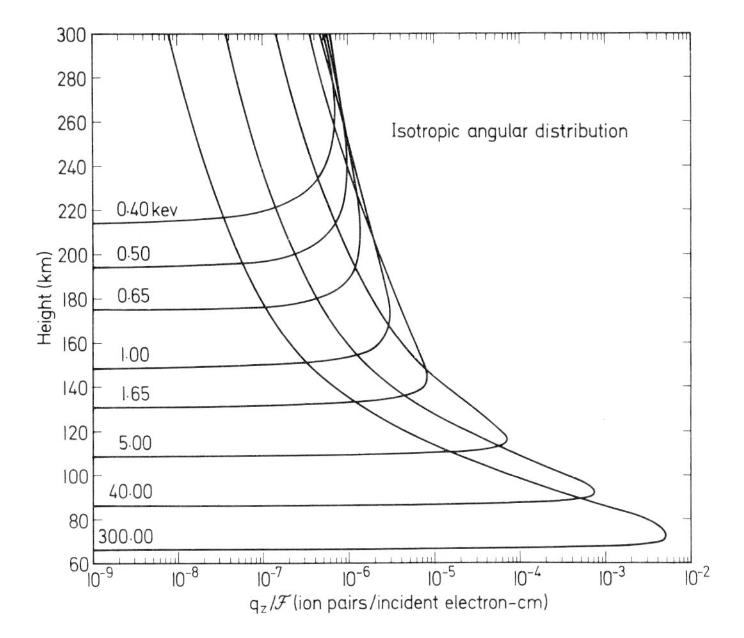

Fig. 2.2 Ionization production by isotropic streams (0°—80°) of mono-energetic electrons. (Rees 1963, courtesy Pergamon Press)

on scattering of primary electrons and on ionisation rate as a function of height have been carried out by Wedde (1970) using Monte-Carlo methods. The ionization profiles agree reasonably well with those of Rees (1963).

2.4 Height Distribution of Other Emissions

For other auroral emissions the relation between the properties of the incident electrons and the emission rate is not as well understood and is therefore known quantitatively to a lesser degree. This problem will be discussed more thoroughly in Sect. 4.2 and in Chapt. 5. As borne out by these discussions, it is likely that the height distribution of the strong, forbidden green oxygen line at 5577 Å usually is similar to that of the first negative N_2^+ bands. Considering the sensitivity curve of the photographic plates used, the intensity distribution in the auroral spectrum, and the similarity in the height distributions of the λ 5577 line and the first negative bands, it is evident also that height distribution curves for auroral luminosity derived from photographic material may give significant information on the energies of the primary electrons. This is important, because a considerable amount of material acquired by photographic techniques is available and may yield considerable information. In recent years, information on height distribution of the luminosity of single auroral lines and bands has begun to emerge, thanks to photoelectric techniques and interference filters.

2.5 Electron Energies Inferred from Height Distribution of Auroral Luminosity

The height of the aurora has been studied by several investigators, most extensively by Størmer (1955, cf. also Egeland and Omholt 1966, 1967). While Størmer and others measured the height of single auroral points which could be identified from several observing points and hence be subject to triangulation, Harang (1946) and Harang and Omholt (1960) studied the luminosity distribution with height in the aurora. All these measurements were made photographically, though, and even if to a certain extent Harang used filters to separate parts of the spectrum

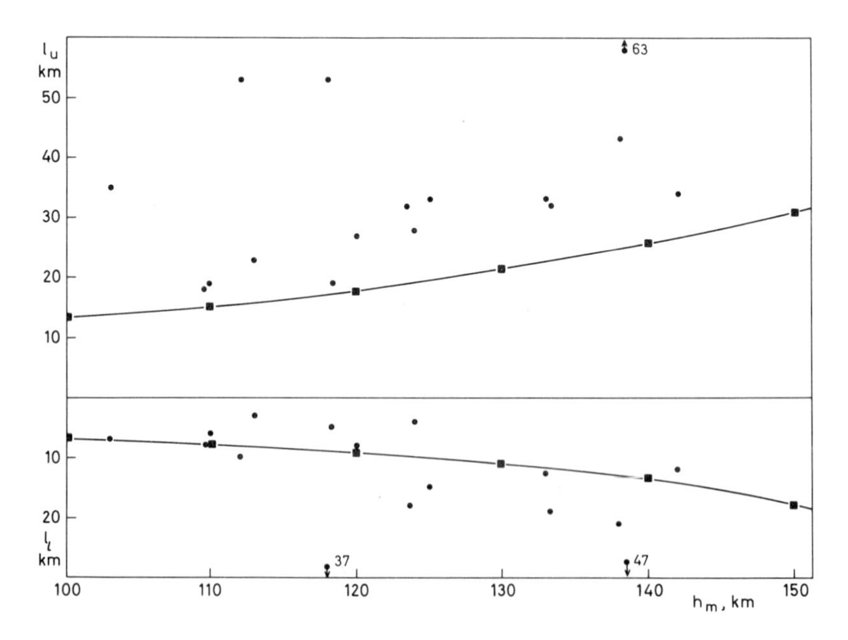

Fig. 2.3 The measured distances from the point of maximum intensity to the upper (l_u) and lower (l_l) points where the intensity has decreased to 50%, as function of the height of the maximum intensity (h_m). ● Belon *et al.* (1966). ■ Harang (1946)

and later a photoelectric technique (Harang 1956), the more recent photoelectric measurements by Belon *et al.* (1966), involving triangulation by photoelectric photometers equipped with narrow interference filters, are certainly superior. We shall therefore consider these first, and then see what additional information can be drawn from older photographic material.

Belon *et al.* (1966) triangulated auroral forms with photometers scanning in the geomagnetic meridian from two stations, College and Fort Yukon. These are located 226 km apart on nearly the same geomagnetic meridian. The measured vertical distances from the point of maximum intensity to the upper (l_u) and lower (l_l) points where the intensity has decreased to half the maximum value, are plotted in Fig. 2.3 as a function of the height of maximum intensity.

Using the basic data on height distribution of luminosity derived by Rees (1963), Belon *et al.* (1966) tried to fit their measured height profiles to an energy spectrum for the primary electrons described by

$$N(E)\,\mathrm{d}E = E^{\gamma} \exp(-E/\alpha) \tag{2.5}$$

and an isotropic angular distribution between 0 and 80°. In general, the best fit was obtained with $\gamma = 1$, although $\gamma = 0$ also gave reasonable results. The best value of α depends on auroral height, being about 10 keV for the lowest aurora and slightly above 1 keV for the highest one. Their data comprise 16 auroral arcs, both homogeneous and rayed. The two auroras which are appreciably different from the others, with maximum luminosity at 118 and 136 km, respectively, are two diffuse homogeneous arcs measured within $1\frac{1}{2}$ minutes of each other. Since the duration of each scan was one minute, it is likely that these data are from the same arc, and that the measurements were influenced by changes in the auroral form during the scan.

The measurements of l_l fit very well into the average curve drawn from Harang's (1946) data, also shown in Fig. 2.3. This curve is derived from measurements of 54 arcs and rayed bands. Above the maximum point, the auroras measured by Belon *et al.* (1966) are apparently more extended than those measured by Harang. Harang did some filter photography and found that the violet part of the spectrum gave a slightly wider height distribution than the green part, but this amounts to only a few kilometers and cannot explain the difference. For l_u, the values found by Belon *et al.* are about $\frac{1}{4}$ to $\frac{1}{3}$ higher than what is consistent with Harang's data, and also fall outside the cluster of Harang's scattered data shown in Fig. 2.4.

There is no obvious explanation for the difference, Harang's data being scattered over more than one solar cycle, without noticeable systematic changes with time. Also, the data are from places with nearly the same L-value.

Måseide (1967) measured the height distribution of the λ 4278 N_2^+ band in two cases of auroral glow by means of rocket-borne photometers. He found a fairly sharp lower boundary at about 95 and 99 km, with maxima at 98 and 102 km. This corresponds to characteristic electron energies α of about 10 keV or slightly more.

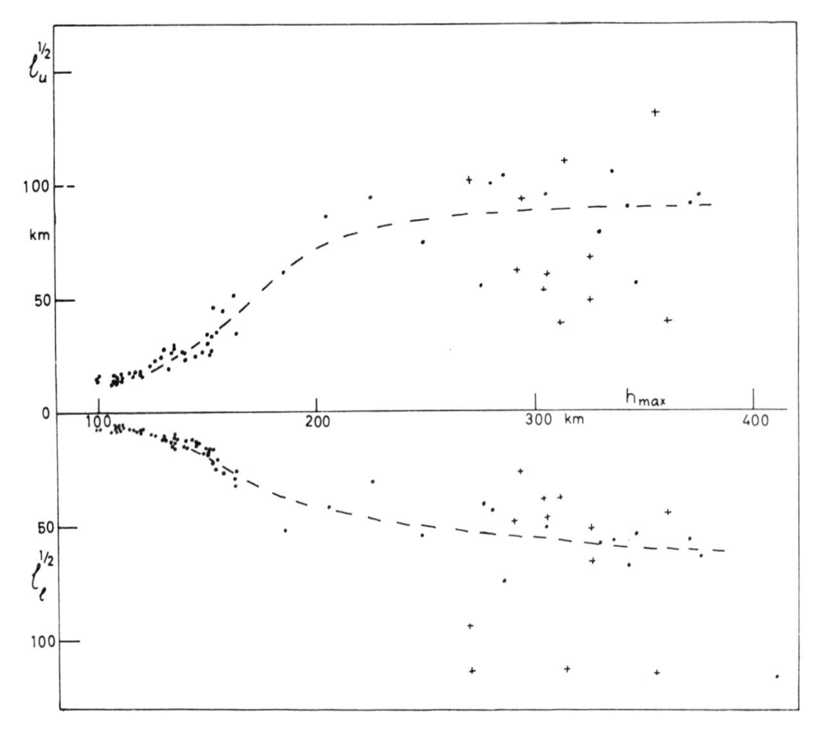

Fig. 2.4 l_u and l_l as function of the height of the maximum intensity. ● auroras in the dark. + sunlit auroras. (From data by Harang 1946 and Harang and Omholt 1960)

Single auroral rays or bundles of rays lie significantly higher in the atmosphere. Harang and Omholt (1960) extended the earlier work by Harang (1946) to greater heights, using Størmer's photographic material. The data are shown in Fig. 2.4. It is evident from Størmer's own measurements (cf. Størmer 1955, Egeland and Omholt 1966, 1967) that single rays are particular in the sense that they show a much greater spread in height and generally lie higher than other forms.

These data also show that there is a slight tendency for some auroral forms, notably homogeneous and rayed arcs, to decrease in height with increasing local time after about 0200 geomagnetic time, but the total shift is less than 10 km. Although Størmer's data are limited to single auroral points, they tend to confirm the conclusion that the characteristic energy for the primary electrons usually varies from about 10 keV for a low aurora to about 1 keV for higher arcs or bands and probably to a few hundred eV for high, single auroral rays. The moderate scatter in the vertical extension of auroral forms with maximum luminosity at a particular height gives a strong indication that the angular and energy distribution of the electrons is a reasonably consistent function of their average energy, with an approximately Maxwellian distribution when the characteristic energy is between 1 and 10 keV.

There is ambiguity in the data, however, since the height distribution is also affected by the distribution in pitch angle of the primary electrons. This ambiguity could be resolved if a reasonable number of ground based optical measurements were correlated with satellite or rocket data. Hence, triangulation from ground may be a powerful tool for studying temporal variations in energy distribution and total particle flux at one site, and provide more useful data related to the detailed studies of the auroral substorm than can be provided by a purely statistical approach.

It is of particular interest to note that long, single rays, which are often fairly stable, are produced by low energy electrons. When rays occur at lower heights, indicating energies of the order 1—10 keV, they most often show rapid movements and changes in structure, being rather unstable. Thus the stability of the striation in the particle beam seems to vary with energy.

There is some evidence for a systematic height variation of auroras with intensity. Data which give evidence for the assumption that intense auroras on the average lie lower in the atmosphere than weaker ones are provided by Harang (1951), Currie (1955), McEwen and Montalbetti (1958) and Starkov (1968). A variation in measured height with intensity of the aurora has also been reported by Kinsey (1965), but it is not quite clear from the observations whether this is merely an effect of sensitivity and contrast. The position of the lower edge is difficult to determine precisely when its intensity is near the threshold of visibility. Kavajiri *et al.* (1965) find a fairly good correlation between the logarithms of the intensities of the λ 5577 and λ 6300 emissions as measured from the ground, but the ratio $I(6300)/I(5577)$ decreases with inereasing intensity. This is another indication that stronger auroras lie lower in the atmosphere than weaker ones (cf. Sect. 4.2.3).

From rapid Doppler temperature measurements made by Hilliard and Shepherd (cf. Sect. 7.2), there is evidence that for a particular aurora the intensity variations are due to variations in the characteristic energy of the primary particles, and to a smaller degree to variations in particle flux. From one aurora to another, however, the relation between the characteristic energy and intensity may differ considerably. Whalen and McDiarmid (1970) found a hardening of the particle spectrum as an instrumented rocket flew into a brighter part of an aurora.

There is also evidence for a decrease in height of the aurora in the course of the night, implying a harder energy spectrum. Such variations found by Størmer (1955) and by Vegard and Krogness (cf. Harang 1951) indicate a decrease in the mean measured height of the lower edge from about 115—120 km three hours after sunset, to about 105—110 km eight hours after sunset. This indicates a systematic change in the characteristic energy by a factor of 2 to 3 in a few hours. The temperature data from optical measurements obtained by Turgeon and Shepherd (cf. Sect. 7.2) show a similar effect, giving a decreasing temperature between 2200 and 0100 hrs. local time, indicating a decrease in average height of the aurora. Radio absorpton data also suggest a hardening of the energy spectrum in the course of the night (cf. sec 8.2). On the other hand, Dzyubenko (1969) finds that the length of auroral rays increases monotonically in the course of the night after 2000 hrs local time.

Finally, we should point out that the emission from sunlit aurora is to a large extent due to resonance scattering of sunlight by N_2^+ ions (cf. Sect. 5.6). Hence it cannot readily be used for estimating the energies of the primary particles.

2.6 Geometry and Motion of the Electron Aurora

Whether the primary particles causing any particular aurora are electrons or protons can be decided from the ground on the basis of two criteria: the absence or presence of hydrogen lines in the spectrum and the spatial structure of the auroral form. As will be pointed out in Chapt. 3, auroras caused by protons have two characteristics: they show Doppler-broadened and -shifted hydrogen lines, and they have a diffuse geometry. The diffusiveness is due to the relatively large proton gyro-radius and to the diffusion of the primary particle stream (cf. Sects. 3.2.2 and 3.2.3).

This does not mean that electrons and protons may not coexist as primary particles; they probably do so to a large extent (cf. Sect. 3.3.5). But those parts of the auroral display which show a fine structure, such as auroral rays and very narrow arcs, are invariably due to electrons, even if protons provide a diffuse background aurora. The gyro-radius of a 10 keV electron in the earth's magnetic field in the auroral zone is only about 6 m or less, depending on pitch angle; that of a proton of the same energy may be as large as 300 m, but the diffusion process described in Sect. 3.2.3 dominates and will wash out any initial structure in a proton stream which is less than ten kilometers or perhaps more.

It is of considerable interest to the theories of the size of the extension of aurora to know the limit for the auroral structures perpendicular to the magnetic field. A lower limit is obviously set by the gyro-radius of the primary electrons, which is of the order 1—10 m. The first attempt to estimate accurately the thickness of narrow auroral arcs yielded a value of about 200 m (Elvey 1957), while Akasofu (1961) obtained values of 150—300 m from structures in an active auroral curtain. Kim and Volkman (1963) measured the overall thickness of auroral arcs in the zenith over Fort Churchill from all-sky cameras and found values ranging from about 3 to 20 km, with an average value of about 9 km.

Because the auroral structures are aligned along the magnetic field, any aurora appearing away from magnetic zenith will be seriously broadened due to perspective effects. Also, observations by photographic techniques are hampered by the movements of the aurora which destroy the fine structure during the finite exposure time of a second or more.

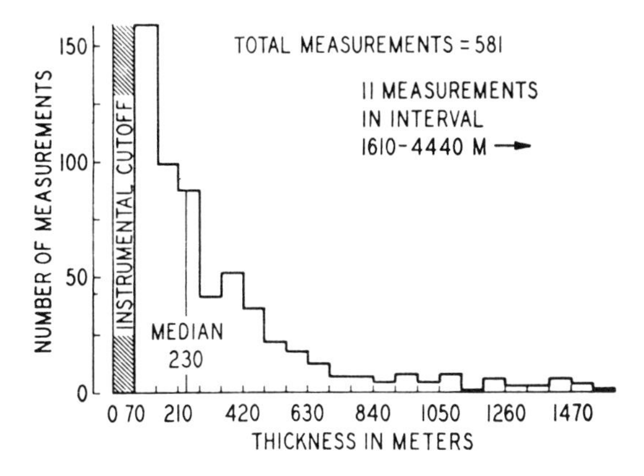

Fig. 2.5 The distribution of thickness of auroral forms. In addition eleven forms were measured in the range 1610—4440 m. (Maggs and Davis 1968, courtesy Pergamon Press)

Realizing this, Maggs and Davis (1968) used a TV image orthicon system with an effective exposure time of 1/60 sec and measured structures of auroras appearing strictly in magnetic zenith. The spatial resolution was such that a structure of 70 m could be detected assuming the altitude of the aurora to be 100 km.

Maggs and Davis claim that auroras observed with such a TV system often appear more rayed and structured than when observed visually, due to the differences in temporal resolution and contrast. Davis (1967) mentions that TV observations often reveal that visually observed homogeneous arcs are composed of parallel, thin arc-like structures, 100 to 300 m thick, which often appear to be streaming horizontally, parallel to the orientation of the overall structure.

Hence, Maggs and Davis measure what they call "thickness of auroral structures" rather than thickness of auroral forms. Fig. 2.5 shows the distribution of the 581 measurements they have made. Another interesting finding was that the structure was narrower the more intense the aurora, and that certain variations with the classification of the overall auroral form occurred. This is shown in Table 2.1.

Table 2.1 *Thicknesses of auroral structures**

Form, type and brightness (IBC: intl. brightness coeff.)	Number	Thickness (in m) Min	Median	Average	Max
All Measurements	581	70	230	380	4440
Homogeneous, IBC 1—2	26	70	422	545	1570
Homogeneous, IBC 2—3	21	70	370	526	1500
Rayed, IBC 2—3	118	70	246	322	1760
Rayed, IBC 3—4	15	82	220	192	1220
Irregular, IBC 1—2	40	70	350	853	4440
Irregular, IBC 2—3	15	75	150	173	262
All Homogeneous	48	70	400	530	1570
All Rayed	135	70	246	320	1760
All Irregular	55	70	262	670	4440
All structures, IBC 1—2	68	70	384	740	4440
All structures, IBC 2—3	154	70	246	335	1760
All structures, IBC 3—4	16	70	220	190	1220

* (from Maggs and Davis (1968)).

An aurora with a sharp geometrical structure often is also moving rapidly. Such movements must be due either to drifts or changes in the source of the precipitating particles, or to changes in the conditions (electric and magnetic fields) experienced by the particles on their way to the ionosphere. This is because the individual primary particles (electrons) give up their energy in a few milliseconds and particles arriving later produce aurora at exactly the same place as long as the position of the source and the path remain unchanged. Exceptions, giving rise to effects which will be discussed later, are the forbidden emissions, with long lifetimes (cf. Sects. 4.1.3 and 5.4).

For a vivid display with rays or during a break-up phase, auroral velocities may be several tens of km s^{-1} (Omholt 1962, Davis 1967), whereas a quiet aurora may show velocities of a few km s^{-1} or less (Chamberlain 1961, Kim and Currie 1960, Stoffregen 1961).

Fig. 2.6 shows the distribution of velocities of auroral rays, measured by a double-slit photometer. The intensity versus time curves measured through two slightly displaced slits shows a time displacement from which the velocity can be determined when a rayed aurora moves by rapidly (Omholt and Berger 1964, also Omholt 1962). In addition to those included in the diagram, a few cases of auroras with extremely high velocities, up to 80 km s^{-1}, were observed.

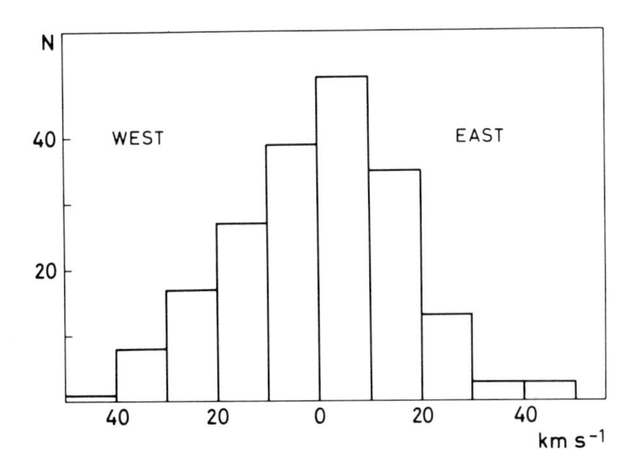

Fig. 2.6 The distribution of drift velocities of rays in aurora (Omholt and Berger 1964)

Danielsen (1969) at Thule measured velocities from 300 to 800 m s^{-1} for auroral forms which moved more than 200 km, while Feldstein *et al.* (1967) measured 130 m s^{-1} for entire forms. This is in agreement with earlier data (Evans 1959).

Hallinan (1969) has measured velocities and wavelengths in the structure near the horizon, using image-orthicon television data. The velocities vary from 0 to approximately 20 km s^{-1} in both east and west directions. There is no apparent typical velocity or systematic relationship with auroral activity. The same velocity ranges and wavelengths (1 to 10 km) were measured in small-scale folds or "curls" in auroral arcs through the magnetic zenith. From this evidence and geometric arguments Hallinan (1969) concluded that the fine ray structure in auroral arcs is produced by the folding of the arc upon itself. This had been suspected for some time, but only high-resolution, coordinated measurements in a direction along, and perpendicular to, the magnetic field lines can provide a final conclusion.

Furthermore, Hallinan (1969) presented strong evidence to explain the small-scale structure of auroral arcs in terms of a sheet beam instability. Instabilities which develop in a sheet beam of electrons impinging on a fluorescent screen show a remarkable resemblance to zenith

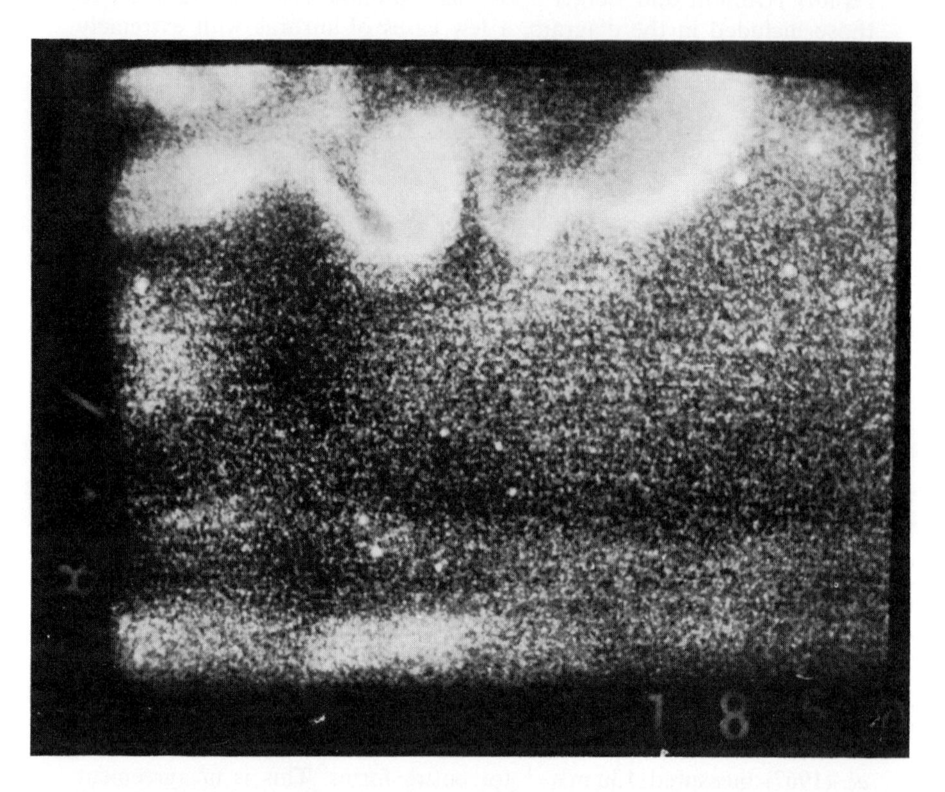

Fig. 2.7 Folds in an auroral arc near magnetic zenith. Distance across photograph $\sim$ 30 km. (Hallinan 1969)

auroras. Laboratory experiments by Kyhl and Webster (1956), Pierce (1956) and Cutler (1956) led Webster (1957) to suggest the sheet beam instablility as a mechanism for the production of auroral rays. He noted that the rotational sense of the fold would indicate the charge of the precipitating particles or the direction of the magnetic field. He noted that the folds in the photographs available to him ran clockwise in the northern hemisphere. Akasofu (1962) and Akasofu and Kimball (1964) also found clockwise folding in photographs from the northern hemisphere. This was confusing because that type of aurora was believed to be produced by electron precipitation.

The use of television in auroral measurements resulted in the observation of small-scale folds in the auroral arc (Hallinan, 1969) with wavelengths of the order of typically 5 km, as shown in Fig. 2.7. 86% of the folds observed in the northern hemisphere ran in the counter-clockwise direction predicted by the sheet beam instability theory. Similarly, most of the events observed in the southern hemisphere were observed to curl in the clockwise direction. Spectral studies showed the aurora to be produced by electrons, while some of the six observed classes of folds were associated with nearby proton precipitation. Thus it is possible that folds in the auroral arcs of the order of 1—10 km can be explained by the sheet beam instability theory. However, large-scale folds and curls (100 km) are apparently less systematic and do not lend themselves to a straightforward explanation in terms of sheet beam instabilities.

Akasofu *et al.* (1969) have discussed a westward travelling surge which occurred during a magnetospheric substorm on March 3, 1968. On the basis of the absence of any detectable Hβ emission and the intensity of the [OI] λ 5577 line, they concluded that the primary particles were mainly electrons, with a flux of 4×10^9 electrons $cm^{-2} s^{-1}$, assuming that the average electron energy was 5 keV. The upper limit for the proton flux was 10^7 protons $cm^{-2} s^{-1}$. On the basis of this and on magnetic records from 70 stations in the northern hemisphere, they analysed the current system in the magnetosphere associated with this substorm. This analysis is an excellent example of how careful and not too elaborate ground-based observations can provide a useful basis for large-scale analysis of the aurora as a magnetospheric phenomenon.

Studies of the striations and movements which arise in artificial ion clouds released in the ionosphere during auroral and quiet conditions may also shed light on the auroral phenomena. Even though the sources of the two phenomena may be at different sites, in the ionosphere and in the magnetosphere, the coupling between these two regions may be strong.

The aurora offers an opportunity, as yet untried, to measure winds and diffusion in the upper atmosphere. As mentioned, the observed movements of auroral forms are not due to atmospheric motions, but to changes in the beam of incident particles, because the emission occurs at the place where the primary particles impinge. However, this is not the case with emissions from metastable atoms or molecules. Because of their relatively long lifetimes (cf. Sect. 5.2), the excited particles may be transported by atmospheric winds through significant distances before emission occurs. For example, if a wind of 100 m s^{-1} crosses an auroral ray, the green oxygen line λ 5577 (emitted by O ^{1}S atoms with average lifetime 0.7 sec, cf. Sect. 5.4) will be emitted with an average displacement of 70 m, and the red oxygen multiplet λ 6300/64 (emitted by O ^{1}D atoms with an average lifetime of 110 sec if sufficiently high up to be undisturbed by collisions, cf. Sects 5.2 and 5.4) will be displaced 11 km. This does not appear as a displacement of the ray as such in green and red light, but as a blurring of the ray to one side, with the intensity falling off with a factor e^{-1} over the distances given above. At great heights, diffusion will also contribute to smear out the rayed structure. A few preliminary measurements by Harang and Fjeld (1964) indicated that the effects of diffusion are indeed measurable on the λ 6300 line. They found that some auroral rays, when viewed in λ 6300 light, appeared roughly one km wider than when viewed in λ 4278 light, and that this result was consistent with present knowledge about diffusion.

Similar measurements of emissions from metastable, excited ions should give the combined effect of wind and diffusion, as well as the Hall drift in the combined electric and magnetic field. A field strength of 0.1 V m^{-1} perpendicular to B gives a drift velocity of about 2 km s^{-1}. Thus information on the electric field could also be obtained from these measurements.

Although they are still rather difficult to perform, with the rapid progress of technique one may anticipate that they might become useful within the foreseeable future.

If the electric field is considerable, an auroral ray should not lie exactly along a magnetic field line. The incident electrons will suffer a drift normal to the magnetic field, in the direction $E \times B$. The deviation of the beam from the magnetic field line is then given by the ratio between the drift velocity and the velocity along the field lines. With an electric field normal to B of 0.1 V m^{-1} the drift velocity is only 20 m s^{-1}, and the angle of deviation becomes negligible, since the primary electrons may have velocities parallel to B of about 10^7 m s^{-1} or higher.

2.7 Latitude Variations in Auroral Heights

Latitude variations occur not only in the general intensity and the total frequency of auroras, but also in the relative distributions among various forms, and in the average heights. Based on more than 12,000 auroral heights measured by Størmer, Egeland and Omholt (1966) found that rays, which generally lie high in the atmosphere, constitute a larger fraction of the forms at high latitudes than at lower ones. The other striking feature in the distribution is the relatively high frequency of pulsating auroras at low latitudes. Also, going from high to low latitudes, there is a shift from rayed arcs and bands to draperies.

This latitude variation in the distribution among different forms is reflected in the average height of the auroral points measured, which increases with increasing geomagnetic latitude. For any particular auroral form the variation is much less pronounced, and is different for different forms (Egeland and Omholt 1966, 1967).

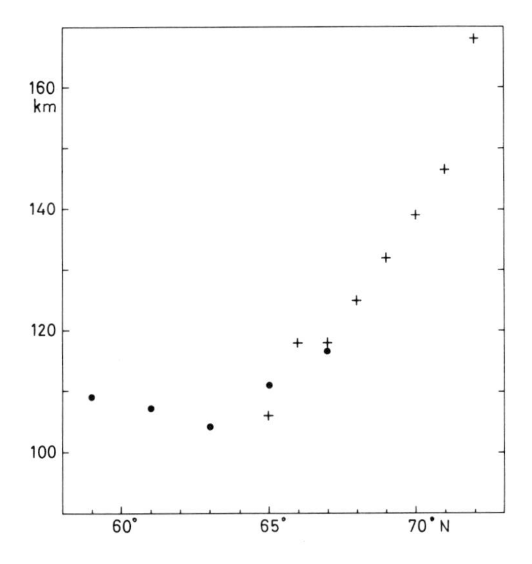

Fig. 2.8 Average heights of quiet auroral arcs in km as function of latitude ● Størmer's data, all measured points (cf. Egeland and Omholt, 1966 or 1967). + Height of maximum luminosity, data by Boyd (1969)

At geomagnetic latitudes greater than about 70°, the increase in auroral heights is more pronounced. Starkov (1968) found that the greater parts of auroras occurring between 75° N and 76° N over Spitzbergen had their lower border in the height interval 145 to 180 km, with a maximum at 175 km, Boyd (1969) measured the height of maximum luminosity in quiet auroral arcs with a triangulation technique. He found a steady increase in the average height from about 120 km at 68° N to 164 km at 72° N. Fig. 2.8 shows his data together with those of Egeland and Omholt (1966, 1967) from Størmer's data. The latter are averages of all points measured in homogeneous arcs, mainly at the lower border.

Andriyenko (1963, 1965a, b, c) has obtained similar results, while Lassen (1969) finds heights around 100—120 km at 75—77° N. Lassen suggests that there may be variations with sunspot activity.

Spectral evidence (cf. Sect. 4.3) also indicates that high auroras are more prominent at high latitudes, inside the auroral oval, than at low latitudes. On the dayside, there is probably a zone of higher, weak auroras equatorwards of the auroral oval, and also inside the oval high auroras are more prominent on the dayside than on the nightside (cf. Sect. 1.2.2).

Some indications have been found for double peaks in the height distribution (cf. Lassen 1969, Boyd 1969), but it is unclear whether these are statistically significant or rather due to observational effects.

References

Aarts, J.F.M., de Heer, F.J., Vroom, D.A.: Physica **40**, 197 (1968).
Akasofu, S.-I.: J. Atmospheric Terrest. Phys. **21**, 287 (1961).
— J. Atmospheric Terrest. Phys. **25**, 163 (1962).
— Eather, R.H., Bradbury, J.N.: Planetary Space Sci. **17**, 1409 (1969).
— Kimball, D.S.: J. Atmospheric Terrest. Phys. **26**, 205 (1964).
Andriyenko, D.A.: Geomagnetizm i Aeronomiya **3**, 913 (1963).
— *Ibid* **5**, 346 (1965a).
— *Ibid* **5**, 684 (1965b).
— *Ibid* **5**, 838 (1965c).
Anger, C.D.: in Aurora and Airglow. (Ed. B. M. McCormac) Reinhold Publ. Co. 1967.
Baker, K.D.: The Birkeland Symposium on Aurora and Magnetic Stroms. (Ed. J. Holtet and A. Egeland) p. 325 1968.
Belon, A.E., Romick, G.J., Rees, M.H.: Planetary Space Sci. **14**, 597 (1966).
Biondi, M.A.: Can. J. Chem. **47**, 1711 (1969).
Boyd, J.S.: Thesis. Univ. of Alaska, College 1969.
Bryant, D.A., Courtier, G.M., Landmark, B., Skovli, G., Lindalen, H.R., Aarsnes, K., Måseide, K.: J. Atmospheric Terrest. Phys. **32**, 1695 (1970).
Chamberlain, J.W.: Physics of the Aurora and Airglow. Academic Press 1961.
Currie, B.W.: Can. J. Phys. **33**, 773 (1955).
Cutler, C.C.: J. Appl. Phys. **27**, 1028 (1956).
Dalgarno, A.: In Atomic and Molecular Processes. (Ed. D.R. Bates) Academic Press 1962.
— Latimer, I.D., McConkey, J.W.: Planetary Space Sci. **13**, 1008 (1965).
Danielsen, C.: Danish Meteor. Inst. Geophys. Rep. **R-9** (1969).
Davidsen, G.: Planetary Space Sci. **14**, 651 (1966).
Davis, T.N.: In Aurora and Airglow. (Ed. B.M. McCormac) Reinhold Publ. Co. 1967.
Dzyubenko, N.I.: Geomagnetizm i Aeronomiya **9**, 290 (Engl. transl.) (1969).
Egeland, A., Omholt, A.: Geofys. Publikasjoner **26**, No 6 (1966).
— — In Aurora and Airglow. (Ed. B.M. McCormac) Reinhold Publ. Co. 1967.
Elvey, C.T.: Proc. Natl. Acad. Sci. U.S. **43**, 63 (1957).
Evans, S.: J. Atmospheric Terrest. Phys. **16**, 191 (1959).
Feldstein, Y.A., Isaco, S.J., Lebedinsky, A.J.: Phenomenology and Morphology and Aurora. U.S.S.R. 1967.
Grün, A.E.: Z. Naturforsch. **12a**, 89 (1957).

Hallinan, T. J.: Thesis. Univ. of Alaska, College 1969.
Harang, L.: Geofys. Publikasjoner **16**, No 6 (1946).
— The Aurorae. Chapman and Hall Ltd. 1951.
— J. Atmospheric Terrst. Phys. **9**, 157 (1956).
— Omholt, A.: Geofys. Publikasjoner **22**, No 2 (1960).
Harang, O., Fjeld, B.: in Studies of Local Morphology, Structure and Dynamics of Aurora. Final Report Contract AF 61–(052)–680 (1964).
Hartman, P. L.: Planetary Space Sci. **16**, 1315 (1968).
— Hoerlin, H.: Bull. Am. Phys. Soc. **7**, 69 (1962).
Hayakagawa, S., Nishimura, H.: J. Geomagn. Geoelect. **16**, 72 (1964).
Jespersen, M., Landmark, B., Måseide, K.: J. Atmospheric Terrest. Phys. **31**, 1251 (1969).
Kawajiri, N., Wahai, N., Nakamura, J., Nakamura, T., Hasegawa, S.: J. Radio. Res. Lab. **12** 141 (1965).
Khare, S. P.: Planetary Space Sci. **17**, 1257 (1969).
Kim, J. S., Currie, B. W.: Can. J. Phys. **38**, 1366 (1960).
— Volkman, R. A.: J. Geophys. Res. **68**, 3187 (1963).
Kinsey, J. H.: J. Geophys. Res. **70**, 579 (1965).
Kyhl, R. L., Webster, H. F.: IRE Trans. Electron Devices **3**, 172 (1956).
Lassen, K.: in Atmospheric Emissions. (Ed. B. M. McCormac and A. Omholt) Van Nostrand Reinhold Co. 1969.
Maggs, J. E., Davis, T. N.: Planetary Space Sci. **16**, 205 (1968).
Mathews, D. L., Clark, T. A.: Can. J. Phys. **46**, 201 (1968).
McConkey, J. W., Latimer, I. D.: Proc. Phys. Soc. **86**, 463 (1965).
— Woolsey, J. M., Burns, D. J.: Planetary Space Sci. **15**, 1332 (1967).
McDiarmid, I. B., Budzinsky, E. E.: Can. J. Phys. **42**, 2048 (1964).
McEwen, D. J., Montalbetti, R.: Can. J. Phys. **36**, 1593 (1958).
McNamara, A. G.: Can. J. Phys. **47**, 1913 (1969).
Måseide, K.: Planetary Space Sci. **15**, 899 (1967).
O'Brien, B. J.: In Aurora and Airglow. (Ed. B. M. McCormac) Reinhold Publ. Co 1967.
Omholt, A.: In Electromagnetic Wave Propagation. (Ed. M. Desirant and J. L. Michiels) Academic Press 1960.
— Planetary Space Sci. **9**, 285 (1962).
— Berger, S.: in Studies of Morphology, Structure and Dynamics of Aurora. Final Report Contract AF 61–(052)–680 (1964).
Philpot, J. L., Hughes, R. H.: Phys. Rev. **133 A**, 107 (1964).
Pierce, J. R.: IRE Trans. Electron Devices **3**, 183 (1956).
Rees, M. H.: Planetary Space Sci. **11**, 1209 (1963).
— Planetary Space Sci. **12**, 722 (1964a).
— Planetary Space Sci. **12**, 1093 (1964b).
— Space Sci. Rev. **10**, 413 (1969).
Sheridan, W. F., Oldenberg and Carleton, N. P.: Abstr. 2nd Int. Conf. Phys. Electronic and Atomic Collisions, Boulder, Colr. 1961.
Spencer, L. V.: Natl. Bur. Std. Monograph No 1 (1959).
Srivastava, B. N., Mirza, I. M.: Phys. Rev. **176**, 137 (1968).
Stadsness, J., Maehlum, B.: Intern Rapport E 053, Norwegian Defence Research Establishment 1965.
Starkov, G. V.: Geomagnetizm i Aeronomiya **8**, 28 (Engl. transl.) (1968).
Stewart, D. T.: Proc. Phys. Soc. **A 69**, 437 (1956).
Stoffregen, W.: J. Atmospheric Terrest. Phys. **21**, 257 (1961).
Størmer, C.: The Polar Aurora. Oxford: Clarendon Press 1955.

Swider, W., Narcisi, R.S.: Planetary Space Sci. **18**, 379 (1970).
Ulwick, J.C.: In Aurora and Airglow. (Ed. B.M. McCormac) Reinhold Publ. Co. 1967.
Walt, M.: In Aurora and Airglow. (Ed. B.M. McCormac) Reinhold Publ. Co. 1967.
Webster, H.F.: J. Appl. Phys. **28**, 1388 (1957).
Wedde, T.: Thesis. Univ. of Oslo. Internal Report E–162. Norw. Defence Research Establishment 1970.
Whalen, B.A., McDiarmid, I.B.: J. Geophys. Res. **75**, 123 (1970).

Chapter 3

The Proton Aurora

3.1 Introduction

The hydrogen lines in the auroral spectrum were first detected by Vegard in 1939 (Vegard 1939a, b). He found that the α (3—2) and β (4—2) lines of the Balmer series (Hα at 6563 Å and $H\beta$ at 4861 Å) occasionally appeared in the spectrum of the aurora, and concluded that these lines were due to showers of hydrogen atoms or protons which sometimes enter the earth's atmosphere during auroras. Later he found that on one occasion the $H\beta$ line was displaced about 5 Å towards shorter wavelengths and interpreted this as due to emission from protons which, while approaching the earth with considerable velocity, were neutralized and excited through encounters with atmospheric atoms and molecules (Vegard 1948).

Meinel (1951) found that the hydrogen lines appeared in spectrograms of auroras in the zenith over Yerkes Observatory, and he was the first to make a detailed study of the profiles these lines, which showed a considerable Doppler shift and broadening. At the same time good spectra, which gave results in agreement with Meinel's observations, were obtained by Gartlein (1950, 1951a, b) and by Vegard and Kvifte (1951).

Fig. 3.1 shows a series of spectra taken with a photoelectric photometer at Kiruna, demonstrating the large Doppler shift and the variability in the intensity of Hα relative to the first positive N_2 bands.

Many observations as well as laboratory investigations and theoretical work have been made in attempts to use the hydrogen emission as a tool to study the impact of protons upon the atmosphere during auroras. Chamberlain (1961) has reviewed work up to 1960, and more recently Eather (1967b) has given a comprehensive review of proton precipitation and hydrogen line emission.

Now that satellites and rockets make direct measurements of the auroral particles possible, the evidence and information which can be obtained from the hydrogen lines may seem superfluous. However, again, the limitations imposed on satellite and rocket measurements

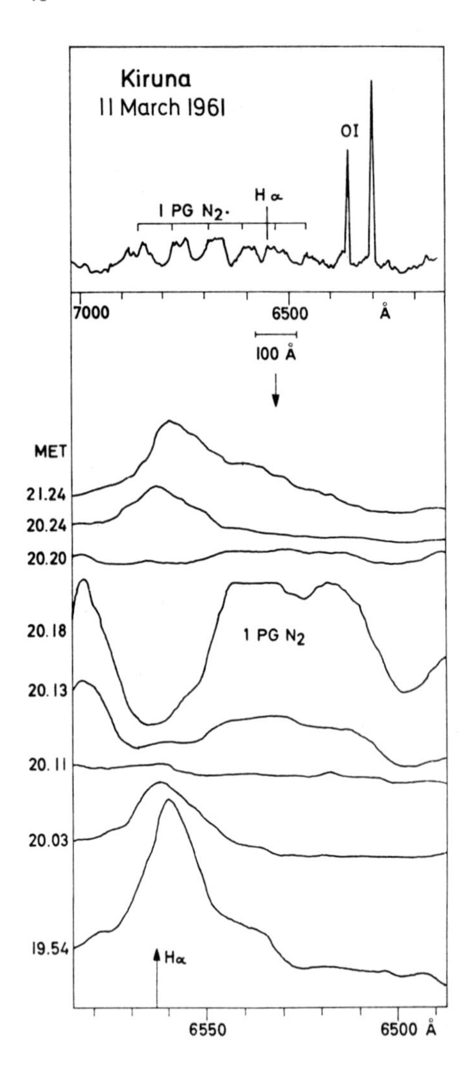

Fig. 3.1 Spectra of aurora in the zenith, showing Hα and first positive bands (Omholt, Stoffregen and Derblom 1962, courtesy Pergamon Press)

by their transient nature make ground-based observations of time and space variations a useful and necessary supplement. Indeed, both kinds of observations are necessary to obtain full information on the proton precipitation during auroras.

In the first part of this chapter (Sect. 3.2) we shall outline the basis for interpretation of the characteristics of the hydrogen emission, such as absolute and relative intensity, Doppler line profile and height distribution. This work is mainly theoretical, based upon laboratory data on basic parameters. Next (Sect. 3.3) we shall see what information can be drawn from observations.

3.2 Theory of Hydrogen Line Emission

3.2.1 Auroral Protons, the Solar Wind and the Magnetosphere

With 100 R as a typical intensity for $H\beta$ during a proton aurora (cf. Table 3.3) the flux of protons required is a few times $10^7\,\mathrm{cm}^{-2}\,\mathrm{s}^{-1}$, perhaps up to $10^8\,\mathrm{cm}^{-2}\,\mathrm{s}^{-1}$, taking the $H\beta$ yield to be a few photons per proton, depending on the energy (cf. Table 3.1). Auroral proton energies are typically of the order of 1—100 keV and their most likely source is the solar wind. From Liouville's theorem it follows that the same intensity of protons as is observed in the aurora must emerge from the region in the magnetosphere where slow protons from the plasma sheet are converted to fast auroral protons, provided the angular intensity distribution is isotropic and the first adiabatic invariant is preserved from the place of acceleration to the atmosphere. By intensity we mean particle flow through a surface perpendicular to the velocity vector; this flow is also often called directional flux. In this book we use the word flux for the flow of particles through a surface perpendicular to the magnetic field direction. The initial source of particles, the solar wind, has a characteristic intensity of 10^8 particles $\mathrm{cm}^{-2}\,\mathrm{s}^{-1}$ (cf. *e.g.* Schardt and Opp 1969). Hence, there appears to be no serious numerical problem regarding the source for auroral protons, although the exact acceleration process is still not known.

Considering in addition the geometrical factors available, Eather (1967b) concludes that much less than one per cent of the protons arriving at the magnetosphere need be converted to auroral protons. He considered the solar wind impacting within a radius of 10 earth radii as a possible source for auroral protons, and took the auroral proton impact area to be two circular zones around the two magnetic poles, located between 60° and 70° latitudes. This gives a geometrical ratio between the source and impact zones of about 300, so that the total source is ample relative to the supply needed. These figures are, of course, indicative only, and they may be considered significant only because they display no obvious numerical disagreement.

Eather (1967b) has pointed out that the kinetic energy density in the proton beam is not negligible compared to that of the static magnetic field. A reasonably strong proton aurora has an intensity of about 100 R in $H\beta$ (cf. Table 3.3). With a characteristic proton energy (e-folding energy of the proton spectrum) of 10 keV as a reasonable working model (cf. Sect. 3.3.6), this requires an influx of about 10^8 protons cm^{-2} s^{-1} (cf. Table 3.1). The total energy influx is then about 10^{12} eV cm^{-2} s^{-1}, or well above 10^{-7} J cm^{-2} s^{-1}*, and requires a proton kinetic energy density in the beam of about 2×10^{-15} J cm^{-3}. If the angular distribution is isotropic all the way out towards the equatorial plane, the energy density here must be 3—4×10^{-15} J cm^{-3} (a factor of 2 higher than that in the influx beam, since in the magnetosphere we must count mirroring protons as well, thus covering a solid angle of altogether 4π instead of 2π). The magnetic induction B is here 1—2×10^{-7} Wb cm^{-2} (1—2×10^{-3} gauss), giving a magnetostatic field energy density of the order of 10^{-14} J cm^{-3}.

Hence, a typical proton aurora is consistent with a moderately perturbed magnetic flux tube out in the magnetic equatorial plane. It is interesting to note that with a very strong electron aurora (several hundreds of kR), assuming excitation by electrons of typically 5—10 keV, one also arrives at an energy density in the beam of a few times 10^{-15} J cm^{-3}. Thus electron auroras are also consistent with a moderately perturbed field.

3.2.2 The Proton Beam in the Atmosphere: Charge Exchange

When protons enter the atmosphere their energy is degraded through inelastic collisions, ionizing and exciting the air molecules and atoms. In some ionization processes the proton may capture the electron lost by the molecule or atom. This process, called charge exchange, is described by

$$H^+ + M \rightarrow H + M^+, \tag{3.1}$$

the hydrogen atom being left either in the ground state or in some excited state. In the latter case the excited hydrogen atom will jump to the ground state or to the metastable $2s$ state by emission, possibly in more than one step. Among the resulting emissions the $H\alpha$ and $H\beta$ lines are possibilities. The neutral hydrogen atom then leaves the magnetic field line, around which the original proton was forced to spiral, and traverses the atmosphere in a straight path until it suffers a new collision, the most likely one being

$$H + M \rightarrow H^+ + M + e. \tag{3.2}$$

* J (Joule) = 10^7 erg = 6.24×10^{18} eV.

Considering a beam of N_+ protons and N_H hydrogen atoms traversing the atmosphere, the change per unit path length in number of protons due to processes (3.1) and (3.2) is given by

$$\frac{dN_+}{dl} = -N_+ N_M \sigma_{10} + N_H N_M \sigma_{01} . \tag{3.3}$$

Here σ_{10} and σ_{01} are the effective cross-sections for processes (3.1) and (3.2) respectively (subscripts 1 and 0 designating state of charge) and N_M is the number density of atmospheric atoms and molecules. Setting $N_M dl = d\zeta$, ζ being the number of atoms and molecules per unit area along the path traversed, and considering that $N_+ + N_H = N$ is constant, we obtain

$$\frac{dF_+}{d\zeta} = - F_+ \sigma_{10} + F_H \sigma_{01} = - \frac{dF_H}{d\zeta}, \tag{3.4}$$

where F_+ and F_H are N_+/N and N_H/N respectively.

If we regard σ_{10} and σ_{01} as constants, we obtain the solution

$$F_+ = F_{+\infty} + (F_{+0} - F_{+\infty}) \exp[-(\zeta - \zeta_0)(\sigma_{01} + \sigma_{10})], \tag{3.5}$$

where

$$F_{+\infty} = \frac{\sigma_{01}}{\sigma_{01} + \sigma_{10}}, \tag{3.6}$$

and F_{+0} represents the value of F_+ for $\zeta = \zeta_0$:

$$F_+ = F_{+0} \text{ for } \zeta = \zeta_0 . \tag{3.7}$$

If the fraction F_+ of protons initially does not have the equilibrium value $F_{+\infty}$, it will approach this value within a distance a few times $\Delta\zeta$, given by

$$\Delta\zeta(\sigma_{01} + \sigma_{10}) = 1 . \tag{3.8}$$

$\Delta\zeta$ is the distance over which the H^+/H particles, on the average, suffer one collision of each kind per particle. The requirement for the integration of eq. (3.4) to be valid is thus essentially that σ_{01} and σ_{10} shall change very little over a few collisions per particle. Since the relevant cross-sections change only slowly with energy, and the fractional energy loss per collision is small, this requirement is indeed fulfilled. Condition (3.8) then shows that only a few collisions per particle bring the composition of the beam (F_+ and F_H) into near-equilibrium. Since the proton/atom bundle suffers a great number of charge-exchange collisions per particle before it is brought to rest, we may consider the composition at all times near the equilibrium value, characterized by $F_+ = F_{+\infty}$.

Since σ_{01} and σ_{10} do change with energy, but slowly, this is also the case with $F_{+\infty}$. Hence, the composition of the beam changes as it traverses the atmosphere and loses energy. The cross-section σ_{10} for charge exchange by process (3.1) is relatively small for very high energies, making $F_{+\infty}$ almost unity. At 100 keV $F_{+\infty}$ is about 0.8 and at 30 keV about 0.5, decreasing to about 0.1 at 3 keV (Eather 1967b).

3.2.3 The Proton Beam in the Atmosphere: Diffusion

The charge exchange of the particles in the proton-hydrogen beam causes an originally narrow beam of protons, spiralling around a magnetic field line as they approach the atmosphere, to diffuse away from it (Omholt 1959). The following rough estimate may illustrate the importance of this effect. Assuming for simplicity a constant scale-height H, and a vertical magnetic field, the proton suffers on average its first charge-exchange collision (3.1) where the atmospheric density is N_1, given by

$$\sigma_{10} N_1 H \sec\theta = 1\,. \tag{3.9}$$

θ is the pitch angle, i.e. the angle between the path of the particle and the magnetic field line We have considered θ to be constant above the height of the first collision. This assumption is too crude for pitch angles close to 90°. $N_1 H \sec\theta$ is the total number of atmospheric particles per unit area, along the path of the proton, from the place where it enters the atmosphere until its first charge-exchange collision.

After being neutralized, the hydrogen atom travels on the average to a point with density N_2, given by the equation

$$\sigma_{01} \cdot (N_2 H \sec\theta - N_1 H \sec\theta) = 1\,, \tag{3.10}$$

before it is re-ionized. Eq. (3.9) and (3.10) yield

$$\frac{N_2}{N_1} = \frac{\sigma_{01} + \sigma_{10}}{\sigma_{01}}\,. \tag{3.11}$$

Denoting with Δh the height difference between the two average collision points, the hydrogen atom travels a distance $\Delta s = \Delta h \tan\theta$ away from the original field line before it is re-ionized and is attached to a new field line. This is the diffusion distance in question, which, again adopting a constant scale height H, is given by

$$\Delta s = \Delta h \tan\theta = H \log_e \left(\frac{N_2}{N_1}\right) \tan\theta\,. \tag{3.12}$$

Using Eq. s. (3.11) and (3.6) we obtain

$$\Delta s = -H \tan\theta \log_e F_{+\infty} . \tag{3.13}$$

The factor $(-\log_e F_{+\infty})$ is small for very high energies, but amounts to about 0.2 at 100 keV, 0.8 at 30 keV and 2.5 at 3 keV, adopting Eather's (1967a, b) data. For an order-of-magnitude estimate, H may be taken to be about 50 km (CIRA 1965 atmosphere, Cospar Working Group 1965). This gives values for Δs in the range 40 to 120 km for particles between 30 and 3 keV, taking $\theta = 45°$. In comparison we may note that the gyro-radii for protons of these energies in the earth's atmosphere are about 500 to 200 m. The diffusion distance decreases rapidly as the particle proceeds downward into the denser atmosphere.

Davidson (1965) has treated this problem in more detail, adopting Monte Carlo methods to study the integrated effect on a thin sheet of protons impinging upon the atmosphere, the geometry being analogous to that of an auroral arc. From his data it appears that the half-width of the arc, due to the diffusion effect described here, is of the order of magnitude of 100 km with 10 keV protons and isotropic angular distribution. This is in reasonable agreement with the values for Δs estimated above, bearing in mind that the first few paths in neutral state will contribute most to the diffusion.

It is unlikely that this effect alone can explain the entire north-south width of hydrogen arcs. However, particles with larger pitch angles θ will suffer the greatest diffusion. Hence, measuring the Doppler shift of $H\alpha$ in the direction along the magnetic field should reveal an edge-effect, $H\alpha$ being sharper and less displaced nearer the northern and southern edges of a hydrogen arc. Furthermore, at large pitch angles an asymmetry in the north-south direction should be seen. Since the magnetic field lines in the auroral zone make an angle of 10° or so with the vertical, hydrogen atoms of pitch angles 80° or more will move upwards when they have a direction close to north and hence escape outwards. Both the diffusion effect and the asymmetry appear to have been observed on the ESRO-I satellite, comparing protons with 80° and 0° pitch angles respectively (Deehr *et al.* 1970).

3.2.4 Excitation of the Hydrogen Lines

The hydrogen atoms in the proton/hydrogen atom beam may be left in any excited state in process (3.1). Hence this process is really composed of a number of processes which are different in detail, and which may be described by

$$\mathrm{H}^+ + \mathrm{M} \rightarrow \mathrm{H}(n\,l) + \mathrm{M}^+ . \tag{3.1 $n\,l$}$$

n and l are the usual orbital and azimuthal quantum numbers. The processes have different cross-sections $\sigma_{10}(n\,l)$ for electron capture into different levels $n\,l$. Another excitation process is direct collision between the hydrogen atoms in the ground state and atmospheric atoms and molecules (M), described by:

$$\begin{aligned} \mathrm{H}(1\,0)+\mathrm{M} \rightarrow\ & \mathrm{H}(n\,l)+\mathrm{M} \\ \text{or}\ & \mathrm{H}(n\,l)+\mathrm{M}^{+}+\mathrm{e}\,, \end{aligned} \tag{3.14}$$

with cross-section $\sigma_{00}(n\,l)$.

The variation of the number $N_0(n\,l)$ of beam particles which are in the $n\,l$ level, is given by:

$$\begin{aligned} \mathrm{d}N_0(n\,l) = & N_0(1\,0)\,N_{\mathrm{M}}\sigma_{00}(n\,l)\,\mathrm{d}s + N_{+}\,N_{\mathrm{M}}\sigma_{10}(n\,l)\,\mathrm{d}s \\ & + \sum_{n'=n+1}^{\infty} \sum_{l'} A_{n'l'nl}\,N_0(n'\,l')\,\mathrm{d}t - \sum_{n'=1}^{n-1} \sum_{l'} A_{nln'l'}\,N_0(n\,l)\,\mathrm{d}t\,, \end{aligned} \tag{3.15}$$

$\mathrm{d}s$ is the distance traversed by the beam during the time interval $\mathrm{d}t$. The two latter terms in the equation represent, respectively, the transitions from higher levels down to the $n\,l$ level and the spontaneous emission from this level to the lower levels. The sums are over all permitted values of l'. A is the corresponding Einstein transition probability.

We have neglected deactivation, i.e. any other cause for transition from excited levels except spontaneous emission. Collisional ionization of excited atoms may be important as a quenching mechanism (Omholt 1959). This effect was investigated in some detail by Bates and Walker (1966) and they concluded that it will not be appreciable in auroras located above 100 km. In PCA-events (Polar Cap Absorption) it may be important considering the large proton energies involved (cf. Sect. 8.6).

The process is obviously important for atoms in the $2s$ configuration. These are metastable, and with no collisional deactivation their number would be substantial, since there is no fast way out by radiation. It is conceivable that collisions are efficient enough to make the number of atoms in this level negligible.

In analogy with the considerations on the equilibrium between protons and hydrogen atoms in the beam, it can be shown that $\mathrm{d}N_0(n\,l)$ in Eq. (3.15) is very small compared to the individual values of the terms on thc right-hand side. Setting $\mathrm{d}N_0(n\,l)=0$ and considering that $\mathrm{d}s/\mathrm{d}t - v$ one obtains from this equation

$$\begin{aligned} I_{nln'l'} = A_{nln'l'}\,N_0(n\,l) = & \frac{A_{nln'l'}}{\sum\limits_{n'=1}^{n-1} \sum\limits_{l'} A_{nln'l'}} \Bigg[N_0(1\,0)\,N_{\mathrm{M}}\sigma_{00}(n\,l)\,v \\ & + N_{+}\,N_{\mathrm{M}}\sigma_{10}(n\,l)\,v + \sum_{n'=n+1}^{\infty} \sum_{l'} I_{n'l'nl} \Bigg]\,, \end{aligned} \tag{3.16}$$

where $I_{nln'l'}$ denotes the number of photons emitted per second from the beam of incident protons.

Dividing I by N and multiplying by $\mathcal{N}_L \mathrm{d}t/N_M \mathrm{d}s$, where $\mathcal{N}_L$ is Loschmidt's number, i.e. the number of atoms and molecules per unit volume of air STP (standard temperature and pressure), one obtains the number J of photons emitted per unit volume of air STP penetrated by a beam of unit area and per particle in the beam (per incident proton):

$$J_{nln'l'} = \frac{A_{nln'l'}}{\sum_{n'=1}^{n-1} \sum_{l'} A_{nln'l'}} \left[F_0(1\,0)\,\sigma_{00}(n\,l)\,\mathcal{N}_L + F_+ \sigma_{10}(n\,l)\,\mathcal{N}_L + \sum_{n'=n+1}^{\infty} J_{n'l'nl} \right], \tag{3.17}$$

where $F_0(1\,0)$ denotes that fraction of the N fast particles which is in the neutral, ground state. The total emission $J_{\mathrm{nn'}}$ in a particular line is obtained by summing over all permitted combinations of l and l'. Because the transition probabilities A also depend on l and l', it is necessary to compute the individual values of J or use proper averages of the values for A and σ. The values of J fall off rapidly with increasing n, and only a few levels above that of particular interest need therefore be considered in the computations.

Eq. (3.17) makes it possible to compute the relevant hydrogen line emission, either per unit air mass penetrated by the particles, or, by use of the known energy-range relation, per unit energy or velocity decrement of the particles.

The current difficulty in the computations is the still limited knowledge of the cross-sections involved. Chamberlain (1954a, b, 1961) and Omholt (1959) circumvented this difficulty by extrapolation (in slightly different ways), using available laboratory and theoretical data on some of the relevant processes as well as data for a proton/hydrogen beam in hydrogen gas. More recently Eather (1966a, 1967a, b) has made new computations based on an improved set of data.

Experimental values for F_{H} and F_+ with reasonable accuracy are available (Allison 1958). For other data, Eather partly based his computations on available information on the charge-exchange processes and partly on extrapolation based on theoretical considerations. However, even these data are ambiguous. For example, different experimenters arrive at values for $\sigma_{10}(4\,\mathrm{s}, \mathrm{p}, \mathrm{d})$ which differ by a factor of 1.6 (Philpot and Hughes 1964, Murray, Young and Sheridan 1966). Hence, even if based partly on experimental data on the processes involved, the excitation of $H\alpha$ and $H\beta$ due to the charge-exchange process is uncertain by an appreciable percentage.

Even more uncertain is the excitation by atom impact given by Eq. (3.14). For this process no direct experimental data are available, and one has to rely on extrapolation from theoretical data for hydrogen impact on hydrogen gas. Even so, the computed values for the hydrogen line emission as a function of particle energy are thought to be correct within a factor 2.

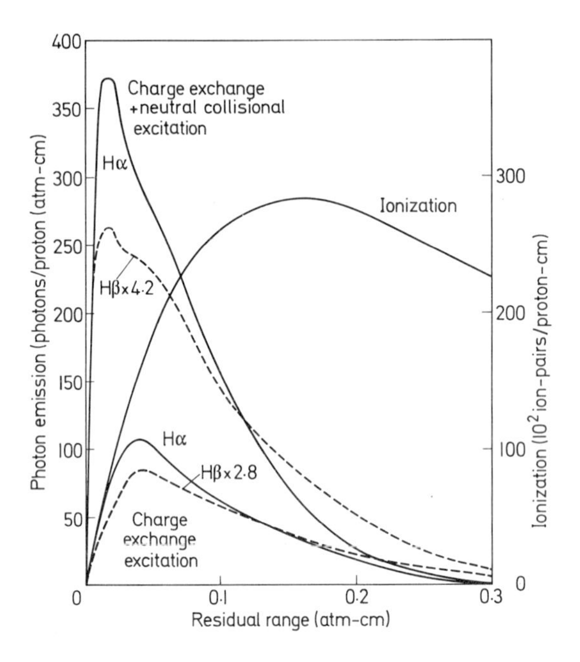

Fig. 3.2 Photon emission curves for Hα and Hβ and ionization curves for protons in air. (Eather 1967b, courtesy American Geophysical Union)

Adopting these data and procedures, Eather arrived at the curves given in Fig. 3.2 for the emission of $H\alpha$ und $H\beta$ photons as a function of energy. The lower curves are based on excitation by charge exchange only, assumed to give a lower limit, whereas the upper curve includes neutral collisional excitation, being relatively more uncertain. For a detailed discussion of the computations, the reader is referred to the review paper by Eather (1967b).

Integrating over the range of the protons, Eather (1967a, b) arrived at the emissions of $H\alpha$ and $H\beta$ photons per incident proton as function of initial proton energy. The results are given in Table 3.1. The Lyman α line should be about 10 times stronger than $H\alpha$ (cf. Eather 1967b).

Table 3.1 *Proton impact on air. Hydrogen line emission*

Initial proton energy keV.	Emitted photons per proton $H\alpha$	$H\beta$
1	≤ 0.4	<0.1
3	1.7	0.3
10	8	1.4
30	17	3.2
100	34	7
300	40	10

3.2.5 The Relation Between Hydrogen Lines and the First Negative N_2^+ Bands

In the earlier literature (cf. Chamberlain 1961, Omholt 1959) it was assumed that the ratio between the total ionization rate and the excitation of the first negative N_2^+ bands was about the same for proton impact as for electron impact. Using the ionization cross-section for protons in air, Eather (1967a, b) arrived at a different result.

However, two processes contribute to the ionization of N_2 by protons:

$$H^+ + N_2 \rightarrow H^+ + N_2^+ + e \tag{3.18}$$

and

$$H^+ + N_2 \rightarrow H + N_2^+ . \tag{3.19}$$

Process (3.18) has a cross-section which increases with proton energy up to about 100 keV (Solovev *et al.* 1962), whereas that of process (3.19) decreases steeply with increasing energy (de Heer *et al.* 1966). Dahlberg *et al.* (1967) and McNeal and Clark (1969) have recently measured the cross-section for excitation of the first negative N_2^+ band by proton impact, and the latter also measured total cross-section for ionization of N_2. It appears that for energies above 10 keV, the ratio between the total rate of ionization and emission of λ 3914 photons is about 30, which is nearly the same as for electrons (cf. Sect. 2.2). For lower energies, this ratio increases rapidly with decreasing energy. At 3 keV the ratio is nearly 10^2 and at 1 keV about 2×10^2.

However, the incident protons will undergo a number of neutralization and re-ionization processes with the result that, if their energy is below 30 keV, they will spend most of their time as hydrogen atoms before being slowed down (cf. Sect. 3.2.2). Therefore, the processes

$$H + N_2 \rightarrow H + N_2^+ + e \tag{3.20}$$

and

$$H + N_2 \rightarrow H^+ + N_2^+ + 2e \tag{3.21}$$

may be as important as processes (3.18) and (3.19). The cross-section for excitation of the first negative bands by hydrogen atom impact was also measured by Dahlberg *et al.* (1967) and by McNeal and Clark (1969). The latter included measurements of the total ionization cross-section for hydrogen impact on nitrogen. The ratio of the rate of ionization to λ 3914 emission is approximately the same as for protons, but somewhat higher at 1 keV ($\sim 4 \times 10^2$).

Table 3.2 gives estimates of the λ 3914 yield per proton and ratios between the λ 3914 and hydrogen line emissions. In arriving at these figures we have used the $H\alpha$ and $H\beta$ yields quoted in Table 3.1, a value of 38 for the ratio between the total rate of ionization of air (including oxygen) and emission of λ 3914, (except at 3 keV, for which a value of 120 was used) and ratios of 1.00, 0.34 and 0.075 as reasonable estimates of the relative photon emissions in the λ 3914, λ 4278, and λ 4709 bands respectively (cf. review of experimental data by Srivastava and Mizra 1968). Further it was assumed that on the average a proton spends about 36 eV per ion pair formed, roughly independent of initial energy (cf. Dalgarno 1962). The data in Table 3.2 must be regarded as rather tentative. Also, due regard should have been given to the fact that secondary electrons will contribute to the ionization and to the excitation of λ 3914. At low energies this will raise the quoted figures because the excitation of λ 3914 will be more efficient than assumed.

Table 3.2 *Proton impact on air. Emission ratios as a function of initial proton energy (theoretical)*

Proton energy	3914 photons per proton	$\frac{3914}{H\alpha}$	$\frac{3914}{H\beta}$	$\frac{4278}{H\beta}$	$\frac{4709}{H\beta}$
3	0.5	0.3	1.5	0.5	0.1
10	7	0.9	5	1.6	0.4
30	20	1.2	6	2	0.5
100	70	2.0	10	3.5	0.8

As stated in Sect. 3.2.4., there are large uncertainties (perhaps up to 50%) in the computed $H\alpha$ and $H\beta$ intensities, because of the uncertainties in the basic cross-sections. Bearing this in mind, the ratios given in Table 3.2 fit well with the observational data obtained by Eather (1968), which yield approximately 14 ± 2 and 1.1 ± 0.1 for the λ 3914/$H\beta$ and λ 4709/$H\beta$ ratio, respectively. About 1 seems to be an observationally lower limit to the λ 4709/$H\beta$ ratio (cf. Sect. 3.3.2).

3.2.6 The Hydrogen Line Profiles

In Sect. 3.2.4 we computed the emission of hydrogen line photons per unit volume of air STP penetrated by a beam of unit area, and per particle ($H^+ - H$) in the beam, given by Eq. (3.17). As stated in that section, this equation permits us to compute the photon emission from any beam of monoenergetic particles as a function of its velocity, which is decreasing as the particles penetrate through the atmosphere. The emission $F(v)$ of photons per decrement of velocity is given by

$$F(v) = J_{nn'}(\xi)\,\frac{\mathrm{d}\xi}{\mathrm{d}v}, \tag{3.22}$$

where ξ is the penetrated air mass at STP. $\frac{\mathrm{d}\xi}{\mathrm{d}v}$ can be obtained from laboratory data of proton penetration through air.

When the photons are emitted, they are each subject to a Doppler displacement given by the component of the particle velocity v in the direction of emission. The purpose of this section is to find the resulting Doppler broadening of the hydrogen lines as function of the initial energy and pitch angle distribution of the protons.

We consider a monoenergetic beam of incident protons with an initial velocity v_0, and an angular distribution $\eta(\theta)$ (proportional to the number of protons per unit time, per unit area through a surface at right angles to the velocity vector, and per steradian; normalized in such a way that $\int_0^{\pi/2} \eta(\theta)\cos\theta\, 2\pi \sin\theta\, \mathrm{d}\theta = 1$). Further, we neglect scattering, so that the pitch angle is constant throughout the atmosphere for any given particle.

The emission of photons from hydrogen atoms with velocity v, within the interval $\mathrm{d}v$, with pitch angle θ contained in a solid angle $\mathrm{d}\Omega$, and per particle in the beam, is then given by:

$$J''(v\,\theta)\,\mathrm{d}v\,\mathrm{d}\Omega = F(v)\,\eta(\theta)\cos\theta\,\mathrm{d}v\,\mathrm{d}\Omega\,. \tag{3.23}$$

Let $\rho(v_x, v_y, v_z)$ be the density of photons emitted by protons with velocity v, described in the velocity space of the emitting particles. Expressed in cartesian coordinates, with the z-axis along the magnetic field, we must have

$$J''(v)\,\mathrm{d}v\,\mathrm{d}\Omega = \rho(v_x, v_y, v_z)\,v^2\,\mathrm{d}v\,\mathrm{d}\Omega\,. \tag{3.24}$$

Using Eq. (3.23) we obtain

$$\rho(v_x, v_y, v_z) = \frac{F(v)\,\eta(\theta)\cos\theta}{v^2}\,. \tag{3.25}$$

The photons are emitted by atoms with different velocities, since the proton/atom beam is degraded in velocity as it penetrates the atmosphere. Those emitters which have the lowest velocity are located at the lowest height, for a given value of θ. Hence the distribution function $\rho(v_x, v_y, v_z)$ varies with height in the atmosphere. In the following we shall neglect this height variation though we in fact consider the emission integrated over height.

An observer looking upwards along the z-axis (the magnetic field line) will observe photons with Doppler displacement corresponding to the velocity component v_z of the atoms from which each is emitted. The emission per particle, per unit v_z and per steradian is $I(v_z)$, given by

$$4\pi I(v_z) = \iint \rho(v_x, v_y, v_z)\, dv_x dv_y . \tag{3.26}$$

The integration is performed over a circular area given by $v_x^2 + v_y^2 = v_o^2 - v_z^2$, where v_0 is the initial velocity of the protons. Outside this area $F(v)$ is zero for the particles in question. The factor 4π on the left-hand side is introduced because the radiation is distributed over a solid angle 4π.

Using Eq. (3.25) and considering the symmetry around the z-axis (since the protons spiral around the field lines with random distribution in azimuth angle), Eq. (3.26) may be written

$$\begin{aligned} 4\pi I(v_z) &= \int_0^{v_{0\perp}} \rho(v_\perp, v_z) \cdot 2\pi v_\perp\, dv_\perp \\ &= 2\pi \int_0^{v_0} \frac{F(v)\eta(\theta)}{v^2} \cos\theta\, v\, dv \\ &= 2\pi v_z \int_0^{v_0} \frac{F(v)\eta(\theta)}{v^2} dv , \end{aligned} \tag{3.26a}$$

where $v_\perp$ is the velocity component at right angles to the z-axis.

The transformation to v is made by making use of the facts that $v^2 = v_z^2 + v_\perp^2$ and that $v_z = v\cos\theta$ is constant during the integration. The latter expression gives the functional dependence of θ on v in Eq. (3.26a).

Similarly, observing in directions at right angles to the magnetic field, the Doppler broadening is described by

$$4\pi I(v_x) = \iint \rho(v_x, v_y, v_x)\, dv_y dv_z , \tag{3.27}$$

the integration area being given by $v_y^2 + v_z^2 = v^2 - v_x^2$. In this case, however, further simplification is not possible.

If, rather than monoenergetic protons, we consider a flux of protons with a relative distribution of initial velocities given by $f(v)$, the function $F(v)$ must be replaced by an integral over initial velocities:

$$\begin{aligned}\mathscr{F}_i(v) &= \int_v^\infty f(v_0) F(v) \, \mathrm{d}v_0 \\ &= F(v) \int_v^\infty f(v_0) \, \mathrm{d}v_0\end{aligned} \tag{3.28}$$

In this case we have implicitly assumed that the angular distribution is independent of the initial energy, i.e. the same for all the incident protons.

From Eqs. (3.26) and (3.27) replacing $F(v)$ by $\mathscr{F}_i(v)$ from eq. (3.28), synthetic zenith and horizon profiles may be computed from given energy and angular distributions of the incident protons. The equations have been generalized for arbitrary observing angles by Tuan (1962) (cf. also Eather 1967b). In Fig. 3.3 are given a number of synthetic $H\alpha$ line profiles computed by Eather and Burrows (1966).

For a first order estimate of the angular distribution, the average Doppler velocities in the zenith and horizon directions are useful (Omholt 1959, cf. also Chamberlain 1961, Eather 1967b). These quantities are

$$\bar{v}_z = \int_0^\infty I_z(v_z) v_z \, \mathrm{d}v_z \Big/ \int_0^\infty I_z(v_z) \, \mathrm{d}v_z \tag{3.29}$$

and

$$\bar{v}_x = \int_0^\infty I_x(v_x) v_x \, \mathrm{d}v_x \Big/ \int_0^\infty I_x(v_x) \, \mathrm{d}v_x . \tag{3.30}$$

For a first order estimate of the angular distribution it is convenient to adopt an analytic form of $\eta(\theta)$. In particular the form

$$\eta(\theta) = \frac{n+1}{2\pi} \cos^n \theta , \tag{3.31}$$

has been used, although without any thorough justification.

Adopting this form of angular distribution, it can be shown that

$$\bar{v}_z = \frac{n+2}{n+3} \frac{\int_0^\infty \mathscr{F}(v) v \, \mathrm{d}v}{\int_0^\infty \mathscr{F}(v) \, \mathrm{d}v} \tag{3.32}$$

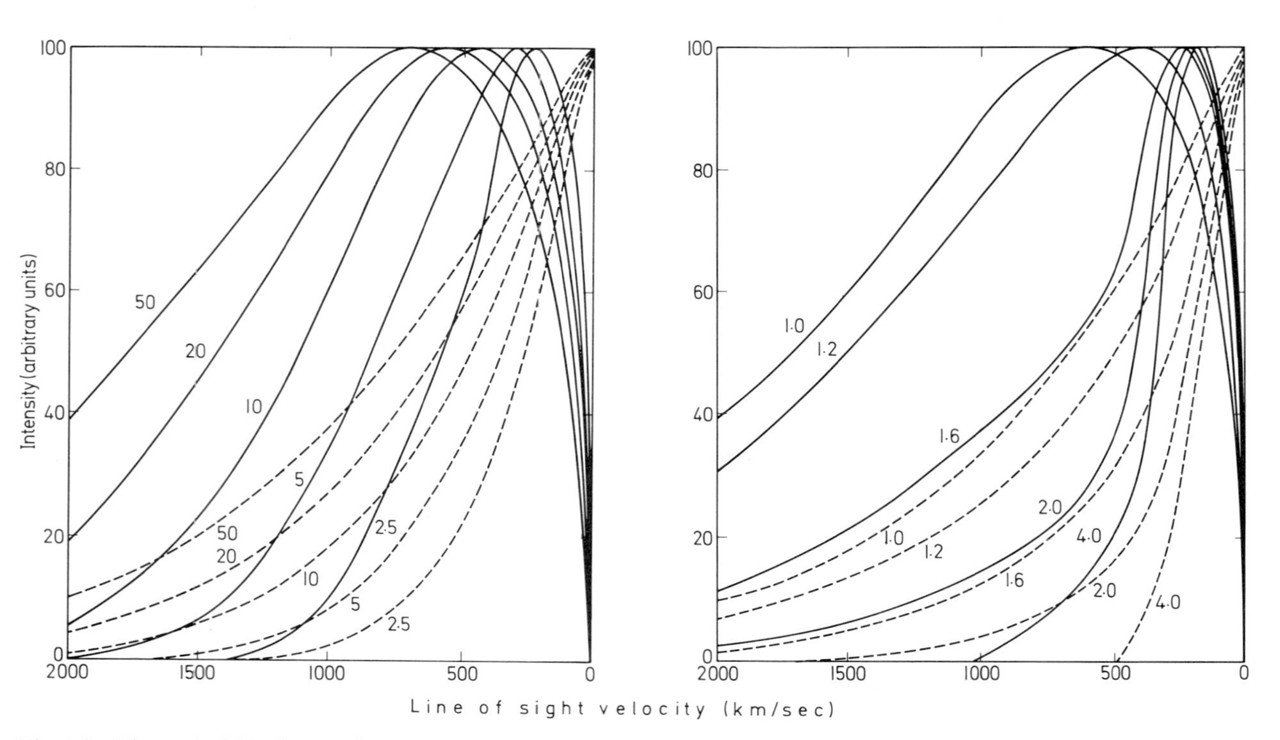

Fig. 3.3 Theoretical Hα line profiles with (left) exponential (exp—E/E_0) and (right) power law ($E^{-\alpha}$) proton energy spectra and isotropic angular distribution. The power law spectra have cutoffs at 1 keV and 1000 keV. Numbers on curves give E_0 and α respectively. Solid lines: zenith profiles. Dashed lines: horizon profiles (the complete horizon profile is symmetric around zero). (Eather and Burrows 1966, courtesy Australian J. Phys.)

and

$$\bar{v}_x = \frac{2(n+2)}{\pi(n+3)} \int_0^{\pi/2} \cos^{n+1}\theta \, d\theta \frac{\int_0^\infty \mathscr{F}(v) v \, dv}{\int_0^\infty \mathscr{F}(v) \, dv}, \tag{3.33}$$

which gives

$$\frac{\bar{v}_x}{\bar{v}_z} = \frac{2}{\pi} \int_0^{\pi/2} \cos^{n+1}\theta \, d\theta . \tag{3.34}$$

Attempts have also been made to use the second order moments ($\overline{v_z^2}$ and $\overline{v_x^2}$) for deriving useful information on the angular distribution, but in this case the tail of the profiles, which are rather inaccurately observed, plays a dominant role. Hence the higher order moments, as determined from observations, are rather uncertain. As will be seen later, the assumption that the angular distribution is independent of the energy of the incident protons is probably incorrect. Nevertheless the ratio $\bar{v}_x/\bar{v}_z$ gives a strong indication whether or not the protons are isotropically distributed ($n=0$), or have a maximum in the forward ($n>0$) or transverse ($n<0$) direction.

With $n=-1$ the flux $(\alpha\eta(\theta)\cos\theta)$ is isotropic, and $\bar{v}_x/\bar{v}_z=1$. In this case the zenith profile must be identical with one half of the horizon profile, since the flux of incident protons is distributed isotropically in all downward directions. However, an isotropic flux is unrealistic, since it would require the intensity $\eta(\theta)$ to be infinite at $\theta=\pi/2$ (cf. Eq. 3.31). Therefore a cut-off or modification in the distribution must be introduced at angles close to $\theta=\pi/2$. With an isotropic directional intensity distribution, which is more likely, $v_x/v_z=2/\pi$.

3.2.7 Height Distribution of the Hydrogen Emissions

In the preceding sections we have considered the emitted light integrated over the height. It is obvious that the emissions are distributed with height in a manner which can be computed, provided the angular distribution, the energy distribution and the emission function $\Phi(\zeta)$ are known with sufficient accuracy. $\Phi(\zeta)$ is the emission per unit air mass penetrated by the particle. Such computations have been made through the years, hopefully with increasing accuracy (cf. Omholt 1956, Chamberlain 1961, Eather and Burrows 1966). The integration is fairly simple as long as one keeps the energy and angular distribution independent of each other. The reader is referred to the book by Chamberlain (1961) and the review paper by Eather (1967b) for details. The height distribution computed by Eather and Burrows (1966) for exponential energy spectra and isotropic angular distribution give height profiles which resemble those for the general auroral luminosity (cf. Figs. 2.3 and 2.4). For characteristic energies between 5 and 50 keV the height of maximum emission ranges from about 130 to 100 km, and the vertical half-width from about 60 to 20 km.

The height distribution of the hydrogen emissions is of less immediate interest than the total intensity and Doppler profiles, because it is rather difficult to measure from the ground. In most cases the hydrogen emission extends over several hundred kilometers in the north-south direction, and thousands of kilometers in the geomagnetic east-west direction. Hence, if at all possible, a rather elaborate triangulation technique by scanning photometers is necessary to deduce the height distribution with significant accuracy. In accordance with this, the scanty measurements from one single observing point give very different results (cf. Eather 1967b).

Rocket techniques could resolve the problem of the height distribution. Comparison between height distribution of the emission and direct measurements of the protons could provide information on the accuracy of the emission function as computed from theoretical and laboratory data. Miller and Shepherd (1968, 1969) have made one flight measuring the height distribution of $H\beta$ in a hydrogen arc. The maximum luminosity occurred at 116 km, with the lower and upper half-intensity points at about 110 and 123 km respectively. Combining the height distribution with direct proton measurements and using accepted cross-sections for $H\beta$ production, they find serious disagreement. A similar experiment by Wax and Bernstein (1970) revealed the same type of disagreement. However, it is probably premature to base conclusions on cross-sections on these single observations. It is obvious, however, that this problem still needs considerable attention.

3.3 Observations on Hydrogen Lines

3.3.1 Techniques

The intensity ratio between the hydrogen lines and the first negative N_2^+ bands allows an estimate to be made of the relative influx of protons and electrons, while the absolute intensity of the hydrogen lines and their Doppler profiles give information about absolute proton fluxes and the energy and angular distribution of the protons. The morphology as well as some properties of the protons can hence be studied through observation of the hydrogen lines.

The earlier studies of hydrogen lines were performed by use of spectrographs, which required long exposure times, and consequently did not yield the necessary time and spatial resolution. The use of photoelectric techniques has made it possible to make vastly more significant observations.

Spectrophotometers, scanning over a limited wavelength range, are greatly superior to spectrographs, and have been extensively used to study the Doppler profile of the hydrogen lines. The first instrument was developed by Hunten (1955) and since then rapid technological development has increased the power of this technique. However, the low light-gathering power inherent in the use of slit spectrographs places severe limits on the use of this instrument.

Photometers with narrow band interference filters have been used since 1957 (Omholt 1957). The recent development of extremely narrow band filters has permitted scanning of the Doppler profile by tilting the filter, a technique which seems extremely powerful (Eather and Jacka 1966). It is now possible to make filters with continuously changing wavelength of the pass-band across the filter or around the edge of a large circular filter. This technique seems particularly promising (Eather 1967 b).

Fabry-Perot spectrophotometers have been used (Zwick and Shepherd 1963), but the sensitivity is inferior to filter techniques. A version which seems more promising is described by Shepherd *et al.* (1965). It allows several points on the line profile to be measured simultaneously.

More recently Francis (1967) has obtained promising results using a narrow band filter with an image-intensifier and obtained valuable data pertaining to morphological problems (Francis and Jacka 1969).

3.3.2 The Intensity Ratio Between the Hydrogen Lines and the First Negative N_2^+ Bands

As is evident from the preceding sections, the intensity ratio between the hydrogen lines and the first negative N_2^+ bands may be used to judge the relative contribution from primary protons and electrons in producing an aurora. Most conveniently, the comparison is made between the $H\beta$ line at 4861 Å and the (0—2) band at 4709 Å, because these are close in wavelength, reducing some of the optical calibration problems. However, the (0—1) band at 4278 Å and (0—0) band at 3914 Å are stronger and hence can be measured with greater absolute accuracy.

A detailed account of earlier measurements of the intensity ratio in question has been given by Eather (1967 b). Different authors have arrived at widely different results. It is likely that this is due to the different morphology of the proton-induced and electron-induced auroras, and the differences in site, observing time and equipment of the different observers. In particular, early observations, say before 1960, suffered from an instrumentation which now is considered to be primitive.

A common lower limit for the λ 4709/$H\beta$ ratio seems to be about unity. The most recent and reliable measurements seem to be those given by Eather (1968) for the ratios λ 3914/$H\beta$ and λ 4709/$H\beta$. These were measurements of what were believed to be typical hydrogen arcs over Fort Churchill (L=8.6) early in the evening (1730—2030 LT), and observed at zenith angles between 60° and 75° to the north, close to the magnetic horizon plane. This time was selected because there was rarely any trace of bright or structured aurora that would indicate electron precipitation (cf. Sect. 3.3.4). A tilting-filter photometer (Eather and Jacka 1966) was used to scan the spectral region of interest, with a band-width of 2—3 Å and a scanning range from 4865 to 4840 Å. In almost all cases the λ 3914/$H\beta$ ratio was between 10 and 17.5, and the λ 4709/$H\beta$ ratio between 0.9 and 1.4. The average values were about 14 ± 2 and 1.1 ± 0.1 respectively, in reasonable agreement with the theoretical data (cf. Sect. 3.2.5, Table 3.2).

From ESRO I satellite photometer data a lower limit of about 3 was found for the λ 4278/$H\beta$ ratio. This is somewhat lower than expected by comparison with the other ratios cited, but in good agreement with the theoretical values given in Table 3.2.

The following conclusion may be drawn from the available data. When the λ 4709/$H\beta$ ratio approaches unity, it is likely that the aurora is produced predominantly by protons. Since large variations in the Doppler profiles are rare (Sect. 3.3.6), the energy and angular distributions of the protons are probably reasonably constant. From this it follows that when the λ 4709/$H\beta$ ratio exceeds unity by a significant amount, that fraction of λ 4709 which provides this higher ratio is due to an electron-induced aurora. The two fractions of the λ 4709 emissions give approximately the relation between the influxes of energy carried into the atmosphere by protons and electrons. The corresponding limits for the λ 4278/$H\beta$ and λ 3914/$H\beta$ ratios are about 3—5 and 10—14 respectively.

3.3.3 Absolute Intensities and Balmer Decrement

Table 3.3 gives some observational data on the $H\alpha$ and $H\beta$ intensities, many of which were compiled by Eather (1967b). Some measurements of the Lyman α line are also included. The Balmer line $H\gamma$ has also been measured (cf. Eather 1967b), but hardly with any significant accuracy.

Taking the rapid development of observational techniques into account, it seems reasonable to place most weight upon recent observations. The conclusion is that intensities of $H\beta$, when present, are typically in the range 10—200 R, occasionally up to 300—400 R, but only rarely above.

Measurements of the intensity ratio $H\alpha/H\beta$ are difficult to make because of the change in detector sensitivity with wavelength. Some data compiled by Eather (1967b) are given in Table 3.4. Considering the experimental difficulties and uncertainties in the theory, these values cannot be considered as inconsistent with the data in Table 3.1, which yield ratios between 4 and 5.5. From Table 3.3 one might guess that 100—150 R is a typical average value for $H\beta$ and about 300 or so is typical for $H\alpha$. This gives a ratio in the range 2—3.

Table 3.3 *Reported hydrogen line intensities*

Author	Line	Intensity
Galperin (1959)	$H\alpha$	Rarely >200 R, though sometimes several thousand R.
Rees and Deehr (1961)	$H\alpha$	Average $\simeq 600$ R and maximum 1.47 kR. Apparently associated with a high-altitude red arc.
Nakamura (1962)	$H\alpha$	Maximum reported 700 R. Average $\simeq 200$ R.
Galperin (1963)	$H\alpha$	Always >100 R.
Eather and Sandford (1966)	$H\alpha$	>150 R near sunspot maximum
Wiens and Vallance Jones (1969)	$H\alpha$	100—200 R in center of hydrogen zone.
Hunten (1955)	$H\beta$	100—1000 R.
Bless and Liller (1957)	$H\beta$	$\simeq 700$ R.
Gartlein and Sprague (1957)	$H\beta$	$\simeq 200$ R.
Omholt (1959)	$H\beta$	20—400 R.
Osterbrook (1960)	$H\beta$	1000 R (during great storms).
Montalbetti and McEwan (1962)	$H\beta$	Maximum observed 600 R. Maxima usually 60—150 R.
Vaysberg (1962)	$H\beta$	$\simeq 100$ R (corrected to the zenith).
Eather and Jacka (1966)	$H\beta$	Maximum observed 150 R. Maximum usually 25—50 R.
Eather (1968)	$H\beta$	Mostly 10—100 R, frequently up to 200 R, occasionally up to 300 R.
Miller and Shepherd (1969)	$H\beta$	60 R.
Francis and Jacka (1969)	$H\beta$	20—30 R in center of hydrogen zone.
Deehr and Sten (ESRO I, private comm.)	$H\beta$	10—50 R Southern part of hydrogen arc 100—200 R Central part, midnight meridian.
Clark and Metzger (1969)	$L\alpha$	40—60 kR.
Hicks and Chubb (OGO 4, Private comm.)	$L\alpha$	Mostly 2—3 kR, maximum 10 kR.

Table 3.4 *Balmer decrement for auroral hydrogen emission*

Author	$I(H\alpha)/I(H\beta)$
Vegard (1956)	7
Shuiskaya (1960)	2.8
	3.0
	3.2
Deehr (1961)	$1.65\,{}^{+0.58}_{-0.34}$

The $L\alpha$ intensity measured by Hicks and Chubb agrees well with the $H\alpha$ intensities quoted and a $L\alpha/H\alpha$ ratio of about 10 (cf. Sect. 3.2.4), while that of Clark and Metzger (1969) is unexpectedly high.

3.3.4 Geometry of Proton Aurora

The most common shape of proton auroras, *i.e.* auroras for which the intensity ratio $\lambda\,4709/H\beta$ is close to the lower limit, is a rather broad arc elongated in the east-west direction (Montalbetti and McEwan 1962, Montbriand and Vallance Jones 1962, Eather and Jacka 1966, Eather

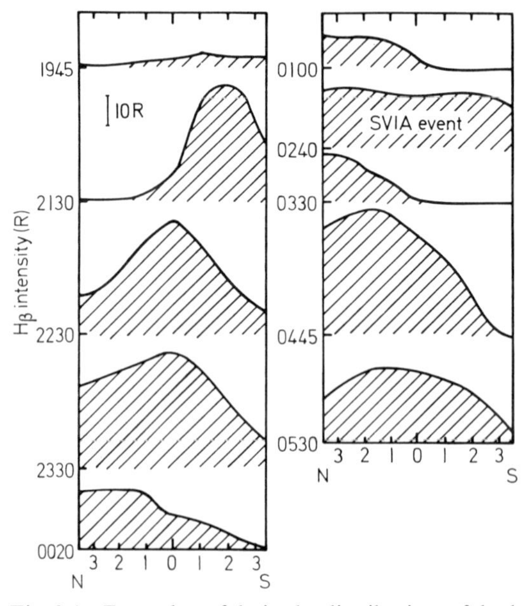

Fig. 3.4 Examples of latitude distribution of hydrogen emission from proton aurora at various local times. The intensities are reduced to zenith (Eather and Sandford 1966, courtesy Australian J. Phys.)

and Sandford 1966, Omholt, Stoffregen and Derblom 1962, Derblom 1968, cf. also review by Eather 1967b). The proton arc is very broad in the north-south direction, the width ranging from about 3 to about 10 degrees of latitude, or 300—1000 km. Typical illustrations of this are given in Fig. 3.4. Also it seems likely that diffuse auroral patches often show strong hydrogen lines, whereas hydrogen emission is not associated with more discrete auroral forms (cf. Omholt 1959, Montbriand and Vallance Jones 1962). The latter result is consistent with theoretical considerations (cf. Sect. 3.2.3).

The relation between proton aurora and electron-induced aurora is rather complicated. In some cases one has an apparently pure proton-excited arc, in other cases there seems to be a mixing of proton and electron influxes. The various observations are partly contradictory and difficult to fit into a simple, consistent picture. A detailed account of the available observations is given by Eather (1967b), and the general morphology of proton aurora is discussed in the next section.

Some observers find that there is a decrease in the hydrogen line intensity when the aurora becomes active; others report an increase in intensity during the break-up phase, associated with a spreading over the sky of the hydrogen emission (cf. Eather 1968). Tsuruda and Kaneda (1968) report from Syowa Base (70° S geom. Lat.) that $H\alpha$ always precedes emission of the first positive bands and auroral break-up.

3.3.5 Morphology of Proton Aurora

Some attempts have been made to construct an auroral hydrogen oval, showing the occurrence and intensity of auroral hydrogen lines in a solar-oriented, geomagnetic coordinate system (cf. the discussion in Sect. 1.2.2). This is difficult, because so much of the relevant observational work has been done by single observers, uncorrelated with other observations. Eather (1967b), on the basis of then available data, constructed an oval-shaped hydrogen zone, centered approximately at 25° from the geomagnetic pole at the day side, and bending strongly inward on the night side. Satellite data (Sharp, Carr and Johnson 1969) on protons are in reasonable agreement with this, but show a double peak on the day side, indicating that the two branches stretching over from the evening and morning side do not overlap or that the oval splits into two parts. Francis and Jacka (1969) made ground-based measurements in Antarctica, on the night side, and also obtained results in agreement with those of Eather (1967b).

Comprehensive studies of hydrogen emission morphology from ground-based measurements have been made by Wiens and Vallance Jones (1969). They studied the hydrogen zone for different magnetic activities, from more than 7500 spectrograms of $H\alpha$ obtained through the years 1964/65. Because of the nature of the observations the data are confined to 12 hours from 1800 to 0600 geomagnetic time. It was found that the center of the proton auroral hydrogen emission oval lies a few degrees equatorwards of the electron (distinct auroras) oval before midnight and crosses to somewhat poleward of it after 0100 hours geom. time, as shown in Fig. 3.5. Derblom (1968) had earlier arrived at similar results. He found that there was a systematic shift between proton and electron auroras, which varied in the course of the night, in the manner described by Wiens and Vallance Jones (1969).

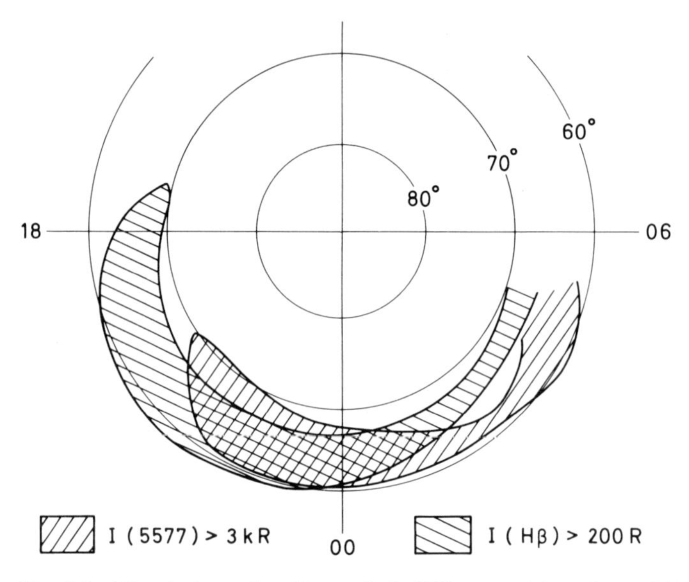

Fig. 3.5 Morphology for Hα and λ 5577 intensities above 200 R and 3 kR respectively during high geomagnetic activity (Kp ≥ 4) (drawn from data by Wiens and Valance Jones 1969)

The average proton and electron precipitation zones overlap considerably. With increasing magnetic activity the hydrogen zone spreads equatorwards and increases considerably in peak intensity. Above geom. lat. 70° the intensity is not very sensitive to geomagnetic activity. For low magnetic activity the hydrogen line intensity is rather symmetrical about magnetic midnight, while for high activity it has a maximum intensity as well as a greater equatorward extent before midnight. Satellite measurements of Lyman α radiation (Clark and Metzger 1969, Chubb and Hicks 1970) and of $H\beta$ (Deehr *et al.* 1970) are in good agreement with ground data. Also, Clark and Metzger's data are consistent with a general conjugacy of proton auroras in the north and in the south. Fig. 3.5 shows the $H\alpha$ and λ 5577 zones for high geomagnetic activity (after Wiens and Vallance Jones 1969). At lower geomagnetic activity the two zones overlap much more, and are closer to the pole.

Rees and Benedict (1970) found that proton auroras always occurred equatorwards of stronger, distinct electron auroras, at all times. This was concluded from observations in Antarctica during 1967. The situation is less clear in the morning hours than in the afternoon, however, and there seems to be a qualitative agreement with Wiens and Vallance Jones' (1969) results, with the proton aurora closer to the electron aurora in the morning hours than at midnight. The different results could perhaps be due to a variation in the size of the proton auroral oval, as compared to the electron auroral oval, with magnetic activity. There seems to be slightly less expansion of the proton oval than of the electron oval with increasing magnetic activity (Deehr *et al.* 1970).

3.3.6 The Doppler Profiles

It is evident from Sect. 3.2.6 that the Doppler profile of the hydrogen lines may be a tool for studying the distribution of protons in pitch angle and energies, but that the full use of this tool is seriously hampered by lack of basic data. But even if the detailed interpretation in terms of accurate pitch angle and energy distributions is still rather dubious, the monitoring of hydrogen line profiles may yield information on systematic and temporal changes in these properties.

Fig. 3.6 shows some observed profiles. It is seen that the profiles are very similar in shape and that all profiles measured in the same direction have about the same half-width. A Doppler velocity of 1000 km s^{-1} corresponds to wavelength displacements of about 22 and 16 Å for the $H\alpha$ and $H\beta$ lines respectively. The horizon profiles shown in Fig. 3.6 have half-widths, in terms of Doppler velocities, of about 700—800 kms^{-1}, and the zenith profiles about 1000 kms^{-1}. The latter, however, show asymmetries with the half-intensity points mostly at about -150 and $+850$ km s^{-1} respectively. The slight asymmetry in some of the horizon profiles may be due to contamination by other emissions. A careful analysis of the contamination is being done by Romick and Stringer (1969).

Most observations support the view that the profiles do not vary drastically from one aurora to another. There are observations, however, which show that occasionally the half-width may vary considerably within a relatively short time (Montalbetti 1959, Galperin 1963, Malville 1960, Johansen and Omholt 1963, Harang and Pettersen 1967).

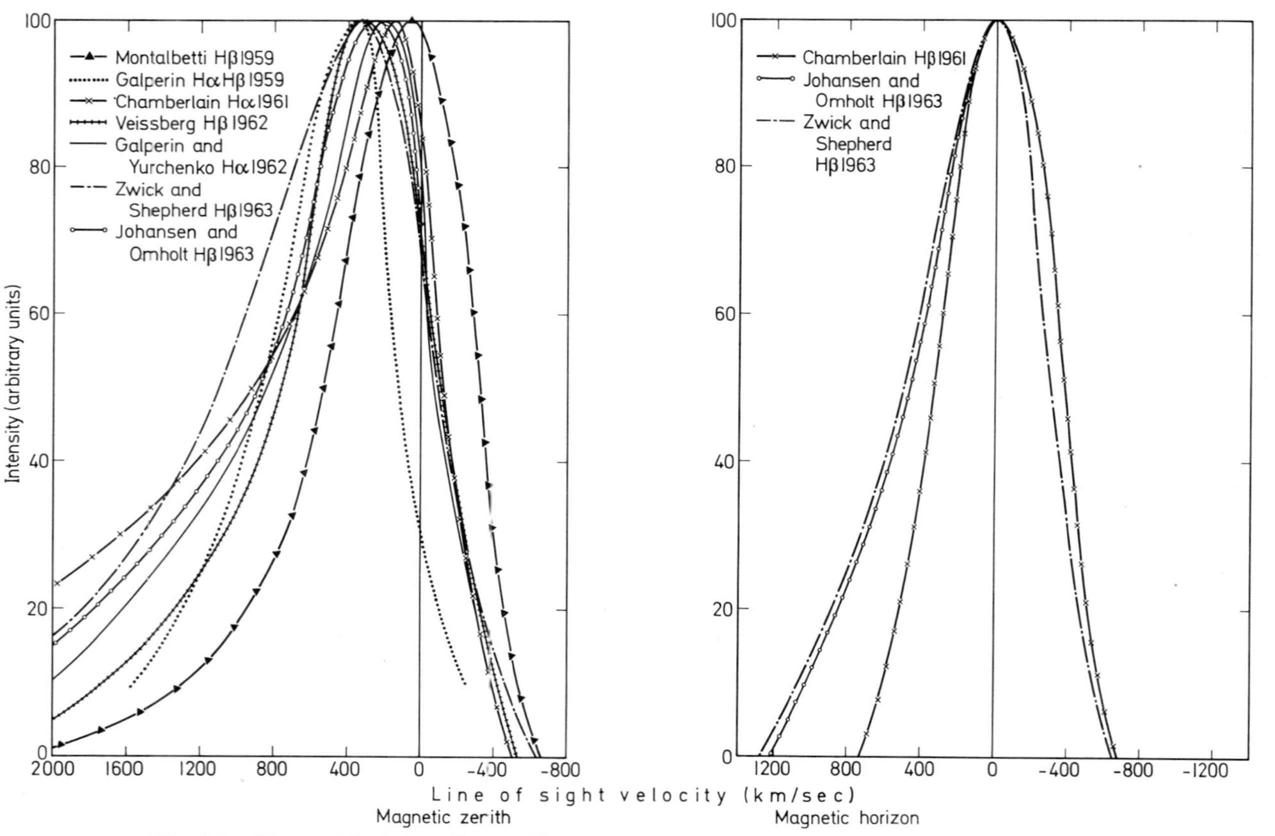

Fig. 3.6 Observed hydrogen line profiles (Eather 1967b, courtesy American Geophysical Union)

The discussion of the observed profiles may conveniently be divided into three parts: i) the main problem of energy and angular distribution, ii) the problem of the observed red-shifted part of the zenith profile, and iii) the variations in width.

i) The general interpretation of the Doppler profiles in terms of energy and angular distribution of the protons is, as mentioned earlier, seriously hampered by inaccuracies in the basic data. Eather (1967b) concludes, from reviewing earlier data, that energy distributions of form $E^{-\alpha}$, with $1.4 < \alpha < 1.8$, or of form $\exp(-E/E_0)$ with E_0 approximately 7 keV, are most consistent with observations. Similarly, assuming a $\cos^n\theta$ pitch angle distribution, the best value of n lies between -1 and $+2$.

These conclusions are based on comparison with theory, and on the assumption that the angular and energy distributions are independent of each other, a rather hypothetical assumption. The derivation of n is in some cases based on the ratio $\bar{v}_x/\bar{v}_z$ given in Eq. (3.34), in others on a more detailed comparison with theory. As pointed out in Sect. 3.2.6, $n=-1$ implies that the zenith profile is identical with one-half of the horizon profile, since then the flux is isotropic, and this is certainly not the case. (The intensity is isotropic for $n=0$, cf. Sect. 3.2.6.) With this model, n is almost certainly zero or positive, the latter possibility being the most likely one (Johansen and Omholt 1963).

Recent satellite data (cf. Deehr *et al.* 1970) indicate that the distribution of pitch angles varies with latitude. A flat distribution (corresponding to negative n) was observed in the equatorward part of an hydrogen arc; these protons may have escaped from the main beam through charge exchange and diffusion processes (cf. Sect. 3.2.3).

A particular difficulty in the interpretation of the profiles is the following: if the parameters are adjusted to make the theoretical zenith profile fit the observations, the theoretical horizon profile shows too long wings and a too sharp maximum at the line center. This contradiction can be eliminated by assuming that high energy protons have a more narrow angular distribution than low energy ones (Omholt 1959, Johansen and Omholt 1963). An angular distribution peaked around $\theta=0^\circ$ for the high-energy protons would reduce the wings in the theoretical horizon profile, and to a certain extent reduce the center of the zenith profile. Conversely, shifting the pitch angles of low energy protons towards larger values would tend to fill in the center of the zenith profile while reducing that of the horizon profile.

ii) The zenith profiles of the hydrogen lines invariably show some emission on the long-wave (red) side of the undisplaced line. The effect is certainly not instrumental, i.e. due to the final instrumental width of a line. Therefore this red-shifted part of the emission must be due to hydrogen atoms which move upward, away from the observer.

Montalbetti (1959) and Zwick and Shepherd (1963) suggested that this red shift is due to photons emitted in other directions being scattered by the atmosphere into the direction of observation. Eather and Jacka (1966), however, rejected this explanation, arguing that the scattering coefficients are not large enough to provide the observed effect. Also, they could not observe any polarization effect which should follow from this mechanism.

Collisional scattering of the protons has also been suggested as important (Bagariatskii 1958a, Chamberlain 1961). However, practically all scattering is expected to be into angles smaller than 0.2°, and even multiple scattering of low energy protons could not explain the observed red shift, which extends to wavelengths corresponding to upward travelling hydrogen atoms with an energy component parallel to the magnetic field of the order 1—3 keV.

Bagariatskii (1958a, b) suggested that the red shift may be due to protons being magnetically reflected, but was unable to reproduce the effect theoretically with the pitch angle distribution adopted by him ($\eta(\theta) \propto \cos^2 \theta$). Eather (1966b), taking into full account the convergence of the magnetic field, concludes that mirroring indeed does provide the necessary effect, provided that the pitch angle distribution for the low energy protons is peaked around 70°—90°, but with an approximately isotropic distribution at lower angles. This conclusion is also compatible with an analysis made by Vaysberg (1966) and qualitatively with the arguments put forward above.

There is as yet little evidence from satellite observations for or against such a distribution occurring persistently. With detectors at 0° and 55° with respect to the magnetic field, Sharp *et al.* (1967) found isotropy within a factor 2, but this is hardly inconsistent with the conclusions from Doppler profile analyses.

iii) Variations in the Doppler width undoubtedly do occur, although such variations usually must be very limited in magnitude. On some occasions, however, rather large and systematic variations occur within limited time intervals.

Exceptionally narrow auroral hydrogen lines have occasionally been reported by several authors (Montalbetti 1959, Malville 1960, Montbriand and Vallance Jones 1966, Harang and Pettersen 1967). Since during one year's recording in Antarctica Eather and Jacka (1966) did not detect any narrow $H\alpha$ emission from aurora, this must be a rare phenomenon. Romick and Stringer (1969) find that narrow $H\beta$ profiles are associated with the break-up phase of auroral displays. A systematic variation of the width of the $H\alpha$ horizon profile in a course of six hours was noted on one occasion by Johansen and Omholt (1963). In that case the profile became wider than usual.

A possible explanation for some of the observed narrow $H\alpha$ profiles may be contamination by emission from the Milky Way (cf. Eather 1967b). Being in certain areas several degrees wide, the latter shows intensities corresponding to 200—600 R. However, rather conclusive evidence for aurorally emitted narrow $H\alpha$ has been obtained by Harang and Pettersen (1967). Near the magnetic horizon they observed a width of less than 5 Å for about one hour, the recorded half-width being about 7 Å with an instrumental one of about 6 Å. Their observations are shown in Fig. 3.7. Comparing these with galactic $H\alpha$ emission data gathered by Montbriand *et al.* (1965), they concluded that galactic emissions could not have contaminated their data.

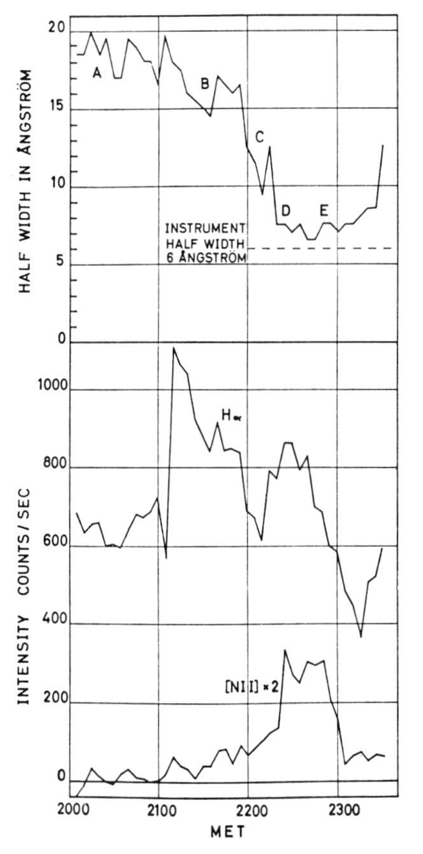

Fig. 3.7 Time variations in Doppler width and intensity of Hα and in the intensity of the [NII] line at 6584 Å (Harang and Pettersen 1967, courtesy Pergamon Press)

Another important feature of their observations is the enhancement of the [NII] line at 6584 Å, which appeared strongly correlated with the narrow $H\alpha$ line. This [NII] line is usually not observed in aurora, probably because the long lifetime of the excited atoms (about 4 min, cf. Table 4.1) makes quenching by collisions predominant. This line was, however, observed in the great red aurora of February 10, 1958 (Belon and Clark, 1959). This aurora occurred at a considerably greater altitude than the average aurora, as demonstrated by the fact that the ratio between the two oxygen line intensities $I(6300)/I(5577)$ was abnormally high (cf. Sect. 4.2.3). Thus, both the narrow $H\alpha$ line and the high intensity of the λ 6584 line observed simultaneously by Harang and Pettersen (1967) are consistent with an aurora at a high altitude, produced by low energy protons.

Even so, it must still be considered as unclear whether the variations in the Doppler profiles are due to variations in energy or pitch angle, or simultaneously in both. Nevertheless, monitoring of $H\alpha$ or $H\beta$ profiles from the ground may still yield valuable information on variations in the properties of the precipitating protons. The value of such observations will be greatly enhanced when the interpretation is better founded than is the case at present.

Certainly, the latitude variation in pitch angle distribution due to diffusion of particles from the main beam (cf. Sect. 3.2.3 and the discussion under i) in this section) may provide an explanation for narrow profiles observed in the zenith.

References

Allison, S. K.: Rev. Mod. Phys. **30,** 1137 (1958).

Bagariatskii, B. A.: Soviet Astron. AJ (English Transl.) **2,** 87 (1958a).

— Soviet Astron. AJ (English Transl.) **2,** 453 (1958b)

Bates, D. R., Walker, J. C. G.: Planetary Space Sci. **14,** 1367 (1966).

Belon, A. E., Clark, K. C.: J. Atmospheric Terrest. Phys. **16,** 220 (1959).

Bless, R.C., Liller, W.: Astronom. J. **62,** 242 (1957).

Chamberlain, J. W.: Astrophys. J. **120,** 360 (1954a).

— Astrophys. J. **120**, 566 (1954b).

— Physics of the Aurora and Airglow. New York: Academic Press 1961.

Chubb, T. A., Hicks, G. T.: J. Geophys. Res. **75,** 1290 (1970).

Clark, M. A., Metzger, P. H.: J. Geophys. Res. **74**, 6257 (1969).

Cospar Working Group IV: Cospar International References Atmosphere. North-Holland, Publ. Co. 1965.

Dahlberg, D. A., Anderson, D. K., Dayton, I. E.: Phys. Rev. **164**, 20 (1967).

Dalgarno, A.: In Atomic and Molecular Processes. (Ed. D. R. Bates) Academic Press 1962.

Davidson, G. T.: J. Geophys. Res. **70,** 1061 (1965).

Deehr, C. S.: Univ. Alaska Sci. Rept. **10** (1961).
— Gustafsson, G. A., Omholt, A., Anderson, L., Egeland, A., Borg, H.: Phys. Norvegica **4**, 101 (1970).
— Sten, T. A., Egeland, A., Omholt, A.: Phys. Norvegica **4**, 95 (1970).
De Heer, F. J., Schutten, J. Maustafa, H.: Physica **32**, 1766 (1966).
Derblom, H.: The Birkeland Symposium on Aurora and Magnetic Storms. (Ed. J. Holtet and A. Egeland) p. 63 (1968).
Eather, R. H.: J. Geophys. Res. **71**, 4133 (1966a).
— J. Geophys. Res. **71**, 5027 (1966b).
— J. Geophys. Res. **72**, 4602 (1967a).
— Rev. Geophys. **5**, 207 (1967b).
— The Birkeland Symposium on Aurora and Magnetic Storms. (Ed. J. Holtet and A. Egeland) 1968.
— In Atmospheric Emission. (Ed. B. M. McCormac and A. Omholt) Van Nostrand Reinhold 1969.
— Burrows, K. M.: Australian J. Phys. **19**, 309 (1966).
— Jacka, F.: Australian J. Phys. **19**, 241 (1966).
— Sandford. B. P.: Australian J. Phys. **19**, 25 (1966).
Francis, R. J.: M. Sc. thesis. Mawson Institute for Antarctis Research, University of Adelaide, South Australia 1967.
— Jacka, F.: J. Atmospheric. Terrest. Phys. **31**, 321 (1969).
Galperin, Yu. I.: Planetary Space Sci. **1**, 57 (1959).
— Planetary Space Sci. **10**, 187 (1963).
— Yurchenko, O.T.: Aurora and Airglow, Results of Researches of the I.G.Y. SSSR. Acad. Sci. **9**, 24—30 (1962).
Gartlein, C. W.: Trans. Am. Geophys. Union **31**, 7 (1950).
— Phys. Rev. **81**, 463 (1951a).
— Nature **167**, 277 (1951b).
— Sprague, G.: J. Geophys. Res. **62**, 521 (1957).
Harang, O., Pettersen, H.: Planetary Space Sci. **15**, 1599 (1967).
Hunten, D. M.: J. Atmospheric Terrest. Phys. **7**, 141 (1955).
Johansen, O. E., Omholt, A.: Planetary Space Sci. **11**, 1223 (1963).
Malville, J. M.: Planetary Space Sci. **2**, 130 (1960).
McNeal, R. J., Clark, D. E.: J. Geophys. Res. **74**, 5065 (1969).
Meinel, A. B.: Astrophys. J. **113**, 50 (1951).
Miller, J. R., Shepherd, G. G.: The Birkeland Symposium on Magnetic Storms. (Ed. J. Holter and A. Egeland) p. 359 (1968).
— — J. Geophys. Res. **74**, 4987 (1969).
Montalbetti, R.: J. Atmospheric Terrest. Phys. **14**, 200 (1959).
— McEwen, D.J.: J. Physical Soc. Japan **17**, Suppl. A-1, 212 (1962).
Montbriand, L. E., Tinsley, B. A., Vallance Jones, A.: Can. J. Phys. **43**, 782 (1965).
— Vallance Jones, A.: Can. J. Phys. **40**, 1401 (1962).
— — Can. J. Phys. **41**, 1393 (1963).
— — Can. J. Phys. **44**, 3259 (1966).
Murray, I. S., Young, S. J., Sheridan, J. R.: Phys. Rev. Letters **16**, C 559 1 (1966).
Nakamura, J.: J. Physical. Soc. Japan **17**, Suppl A-1, 227 (1962).
Omholt, A.: J. Atmospheric Terrest. Phys. **9**, 18 (1956).
— Astrophys. J. **126**, 461 (1957).
— Geofys. Publikasjoner **20**, No. 11 (1959).
— Stoffregen, W., Derblom, H.: J. Atmospheric Terrest. Phys. **24**, 203 (1962).
Osterbrook, D. E.: Science **131**, 353 (1960).
Philpot, J. L., Hughes, R. H.: Phys. Rev. **133**, 107 (1964).

Rees, M. H., Benedict, P. C.: J. Geophys. Res. **75**, 1763 (1970).
— Deehr, C. S.: Planetary Space Sci. **8,** 49 (1961).
Romick, G. J., Stringer, W. J.: In Annual Report, Geophysical Institute, University of Alaska 1969.
Sandford, B. P.: J. Atmospheric Terrest. Phys. **26,** 749 (1964).
Schardt, A. W., Opp, A. G.: Rev. Geophys. **7,** 799 (1969).
Sharp, R. D., Carr, D. L., Johnson, R. G.: J. Geophys. Res. **74,** 4618 (1969).
— Shea, M. F., Shook, G. B., Johnson, R. G.: J. Geophys. Res. **72,** 227 (1967).
Shepherd, G. G., Lake, C. W., Miller, J. R., Cogger, L. L.: Appl. Opt. **4,** 267 (1965).
Shuyskaya, F. A.: Soviet Astron. AJ **4,** 178 (1960).
Solov'ev, E. S., Il'in, R. N., Oparin, V. A., Fedorenko, N. V.: Soviet Phys. JETP (english transl.) **15**, 459 (1962).
Srivastava, B. N., Mirza, I. M.: Phys. Rev. **176,** 137 (1968).
Stoffregen, W., Derblom, H.: Planetary Space Sci. **9,** 711 (1962).
Tsuruda, M., Kaneda, E.: Rep. Ionosph. Space Res. Japan **22,** 289 (1968).
Tuan, R. F.: Astrophys. J. **136,** 283 (1962).
Vaysberg, O. L.: Aurora and Airglow, Results of Researches of the I. G. Y. SSSR. Acad. Sci. **8,** 36 (1962).
— Geomagnetizm i Aeronomiya **6,** 101 (1966).
Vegard, L.: Nature **144,** 1089 (1939a).
— Geofys. Publikasjoner **12,** No. 14 (1939b).
— Proc. I. U. G. G. Conf. in Oslo 1948.
— Geofys. Publikasjoner, **19,** No. 9 (1956).
— Geofys. Publikasjoner **18,** No. 3 (1951).
Wax, R. L., Bernstein, W.: J. Geophys. Res. **75,** 783 (1970).
Wiens, R. H., Vallance Jones, A.: Can. J. Phys. **47,** 1493 (1969).
Yevlashin, L. S.: Geomagnetizm i Aeronomiya **3,** 405 (1963).
Zwick, H. H., Shepherd, G. G.: J. Atmospheric Terrest. Phys. **25,** 604 (1963).

Chapter 4

The Optical Spectrum of Aurora

4.1 Description

4.1.1 Introduction

The optical spectrum of aurora is characterized by numerous emission lines and bands from atomic and molecular nitrogen and oxygen, neutral as well as singly ionized. In addition, lines from sodium and helium are occasionally observed, and hydrogen lines are regular features in proton aurora. More or less successful attempts to identify the various lines and bands in the auroral spectrum have been made for almost a century. This work took a great upswing about 1950, when improved spectrographs became available. A detailed description of the optical spectrum at wavelengths from about 3300 Å to 11000 Å was given by Chamberlain (1961). Since then, little new information has emerged about this wavelength region. Rocket and satellite data, however, have extended our knowledge about the spectrum into the ultra-violet region, and some more information has become available in the infra-red.

For a physical interpretation of the spectrum in terms of excitation mechanisms, intensities are as important as identification of the lines and bands. Intensity measurements are, however, in general very difficult to make with high accuracy, due to the delicate calibration technique. Another difficulty with intensity measurements is that the geometry of the aurora often precludes a proper interpretation. Any instrument observing auroral light necessarily integrates all the light emitted within a volume of space covered by the instrument's field of view along the line of sight of the instrument. Therefore it is difficult, and often impossible, to correlate intensities to particular points within the aurora. For these reasons, intensity measurements of weak emissions are very poor, and only indicative, while those of stronger emissions are still uncertain to an uncomfortable degree.

In this description of the auroral spectrum, we shall limit ourselves to what seems physically relevant. For example, whether a particular permitted oxygen or nitrogen line is observed with a particular intensity often is not of great interest. What is relevant, however, is that the permitted lines in the spectrum in general show that atomic oxygen and nitrogen are excited to the lowest-lying terms.

4.1.2 The Auroral Spectrum

The optical spectrum of aurora observed hitherto is shown in Figs. 4.1 to 4.7. These figures include identifications of the most important atomic lines and molecular bands.

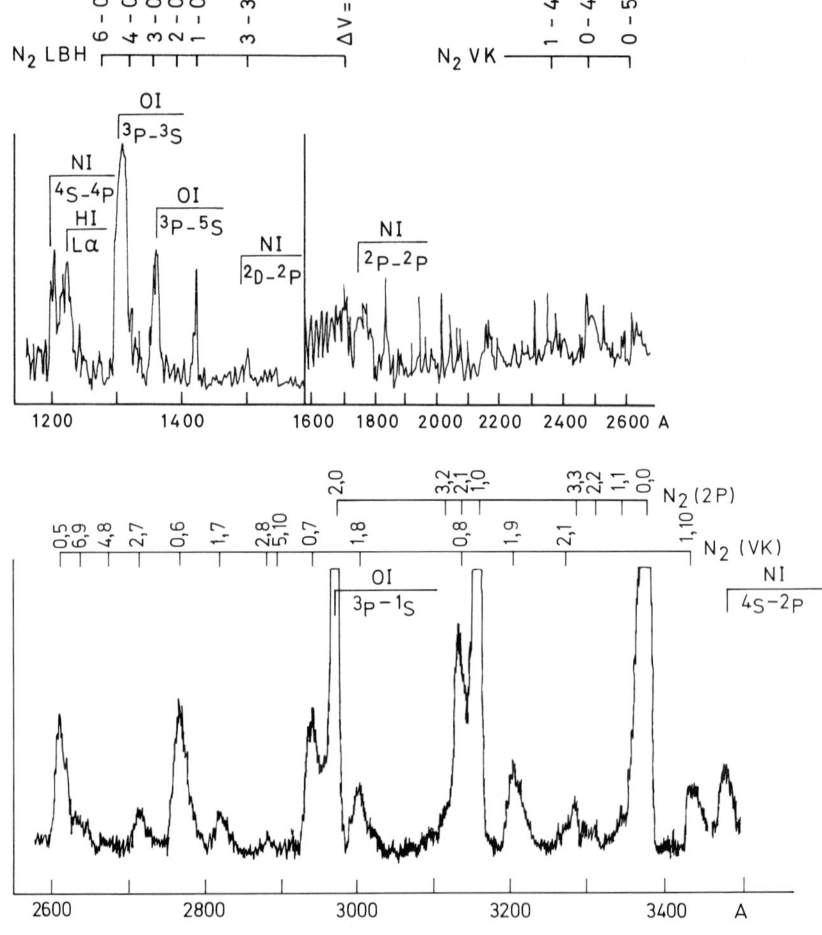

Fig. 4.1 The ultraviolet spectrum of aurora. The spectra were kindly supplied by C. A. Barth, M. H. Rees, and W. E. Sharp. The permitted OI and NI Lines belong to the 2p—3s transitions. Note change in wavelength scale at about 1600 Å

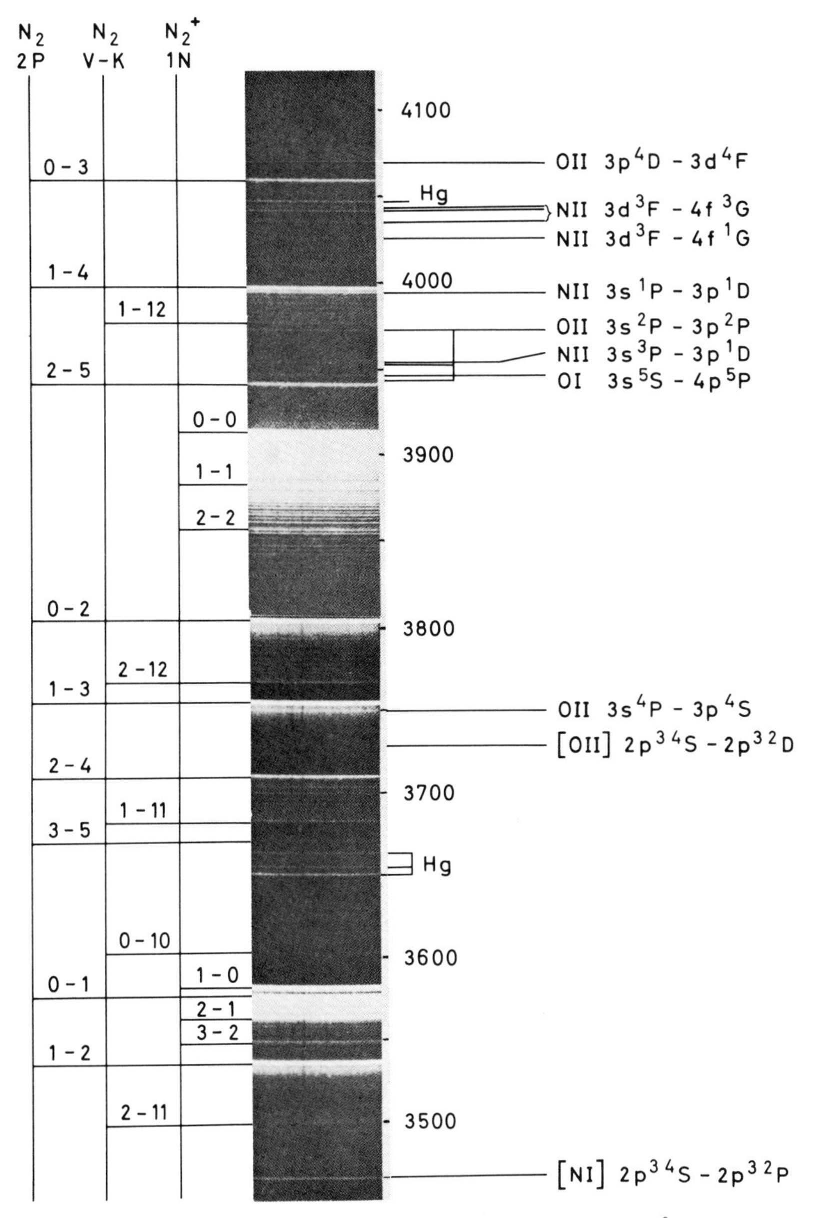

Figs. 4.2—4.6 The auroral spectrum between 3400 and 8800 Å. The spectra were kindly supplied by A. Vallance Jones. The identifications shown are all considered to be certain. A few lines and bands which do not appear in these particular spectra are included among the identifications, and their positions are indicated. (Source for identifications: Chamberlain 1961, Vallance Jones 1969, Remy, Arpigny and Rosen 1960)

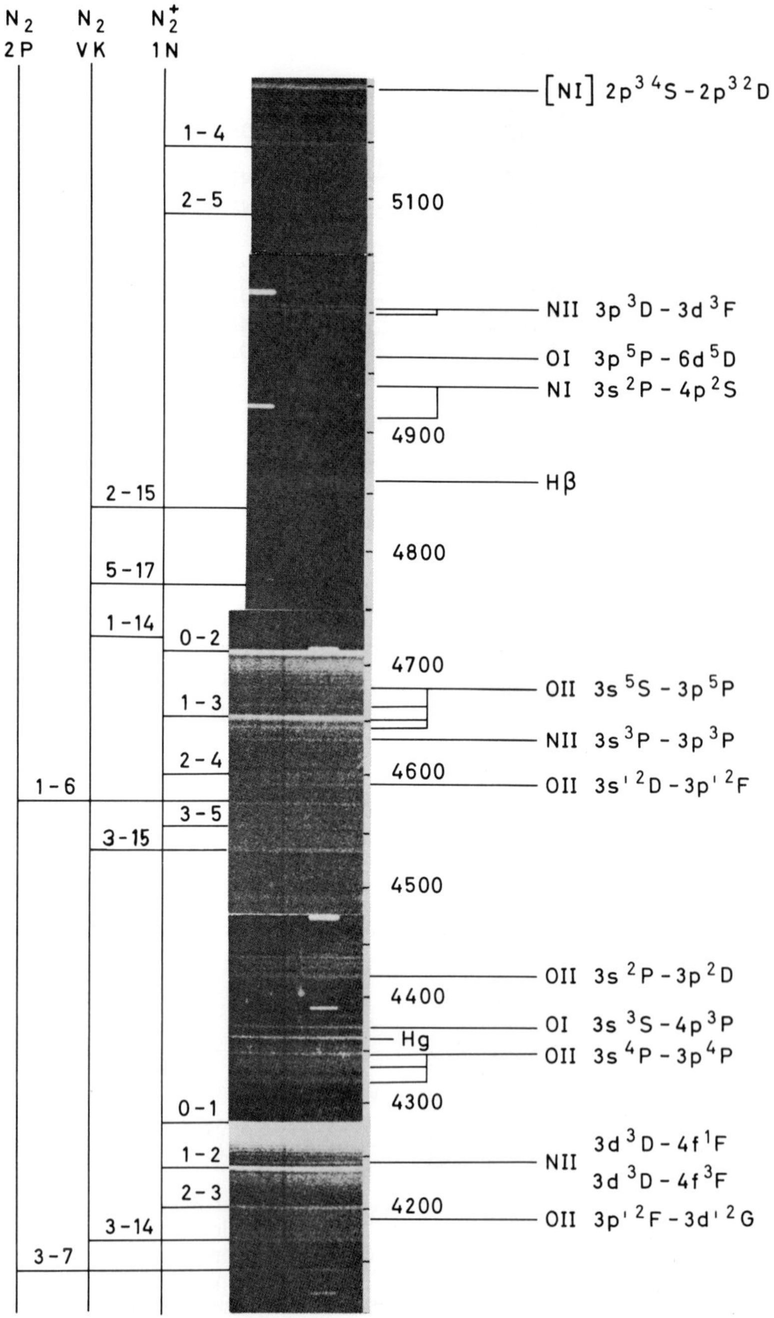
N2 2P
N2 VK
N2+ 1N
1-4
2-5
2-15
5-17
1-14
0-2
1-3
2-4
1-6
3-5
3-15
0-1
1-2
2-3
3-14
3-7
[NI] 2p3 4S - 2p3 2D
5100
NII 3p 3D - 3d 3F
OI 3p 5P - 6d 5D
NI 3s 2P - 4p 2S
4900
Hβ
4800
4700
OII 3s 5S - 3p 5P
NII 3s 3P - 3p 3P
4600
OII 3s' 2D - 3p' 2F
4500
OII 3s 2P - 3p 2D
4400
OI 3s 3S - 4p 3P
Hg
OII 3s 4P - 3p 4P
4300
NII
3d 3D - 4f 1F
3d 3D - 4f 3F
4200
OII 3p' 2F - 3d' 2G

Fig. 4.3

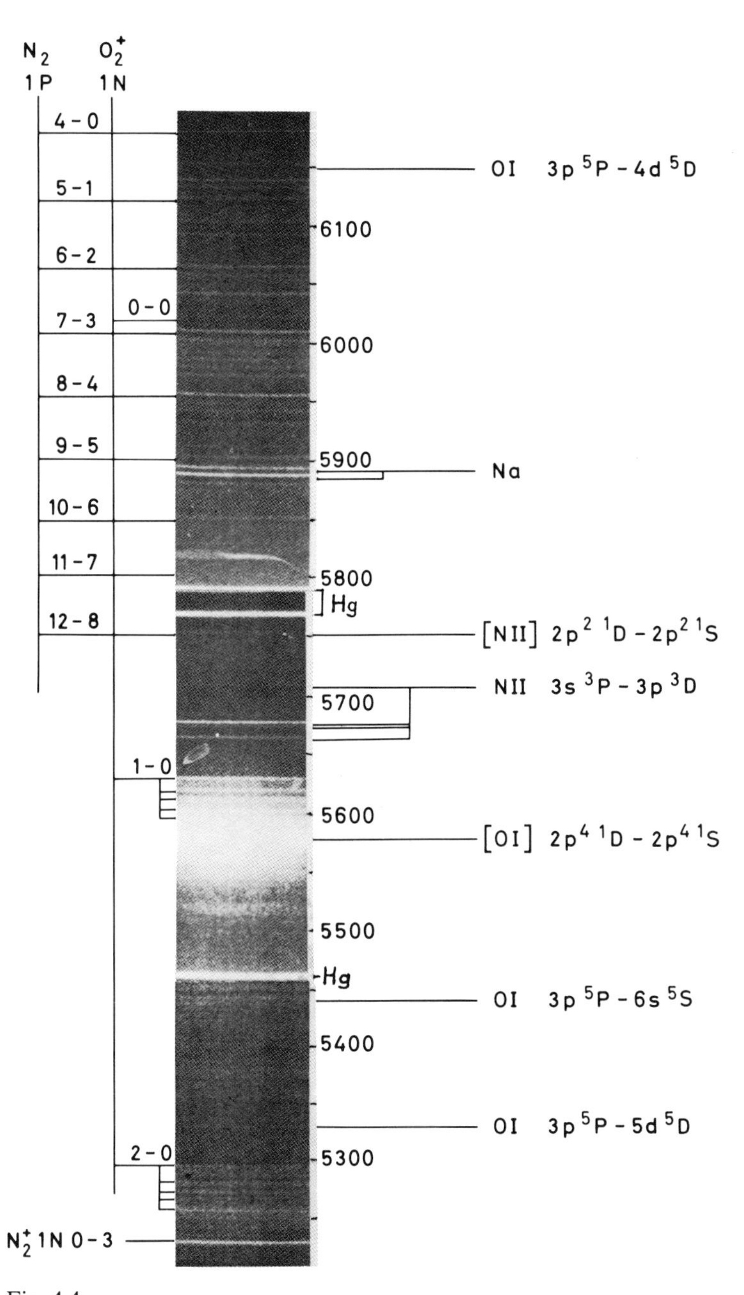

N2 1P
O2+ 1N
4-0
5-1
6-2
7-3
0-0
8-4
9-5
10-6
11-7
12-8
1-0
2-0
N2+ 1N 0-3
6100
6000
5900
5800
5700
5600
5500
5400
5300
Hg
Hg
OI 3p 5P-4d 5D
Na
[NII] 2p2 1D-2p2 1S
NII 3s 3P-3p 3D
[OI] 2p4 1D-2p4 1S
OI 3p 5P-6s 5S
OI 3p 5P-5d 5D

Fig. 4.4

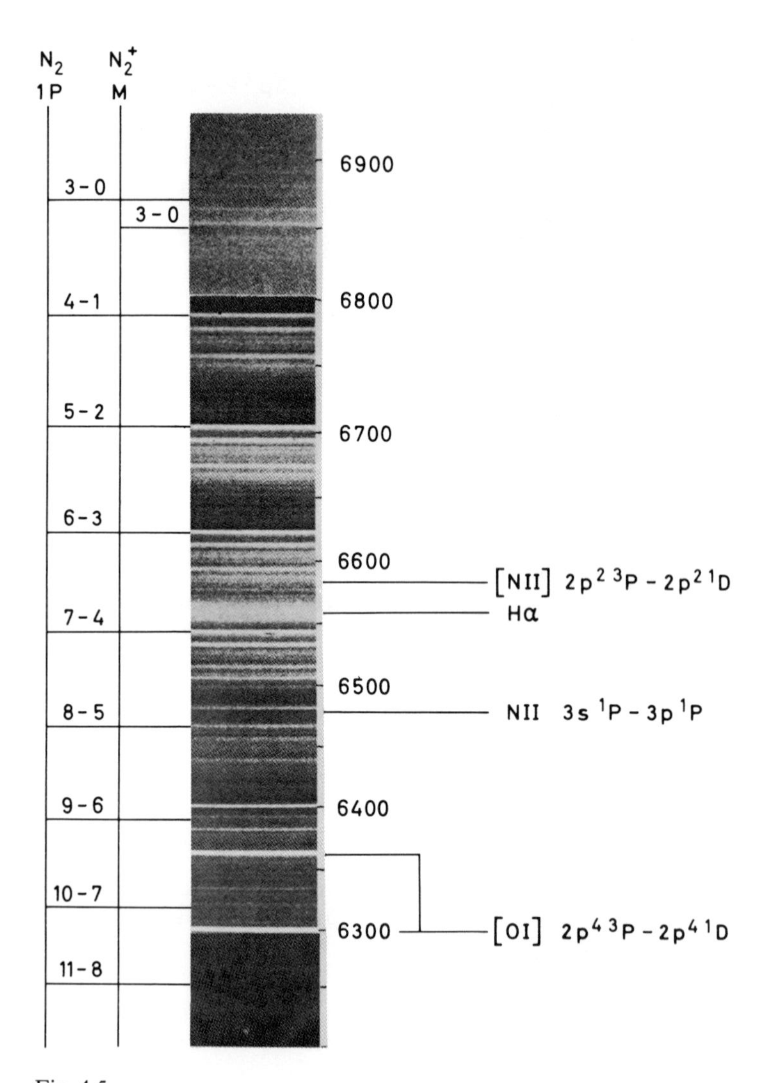
N2
1P
N2+
M
3 - 0
3 - 0
4 - 1
5 - 2
6 - 3
7 - 4
8 - 5
9 - 6
10 - 7
11 - 8
6900
6800
6700
6600
6500
6400
6300
[NII] 2p2 3P - 2p2 1D
Hα
NII 3s 1P - 3p 1P
[OI] 2p4 3P - 2p4 1D

Fig. 4.5

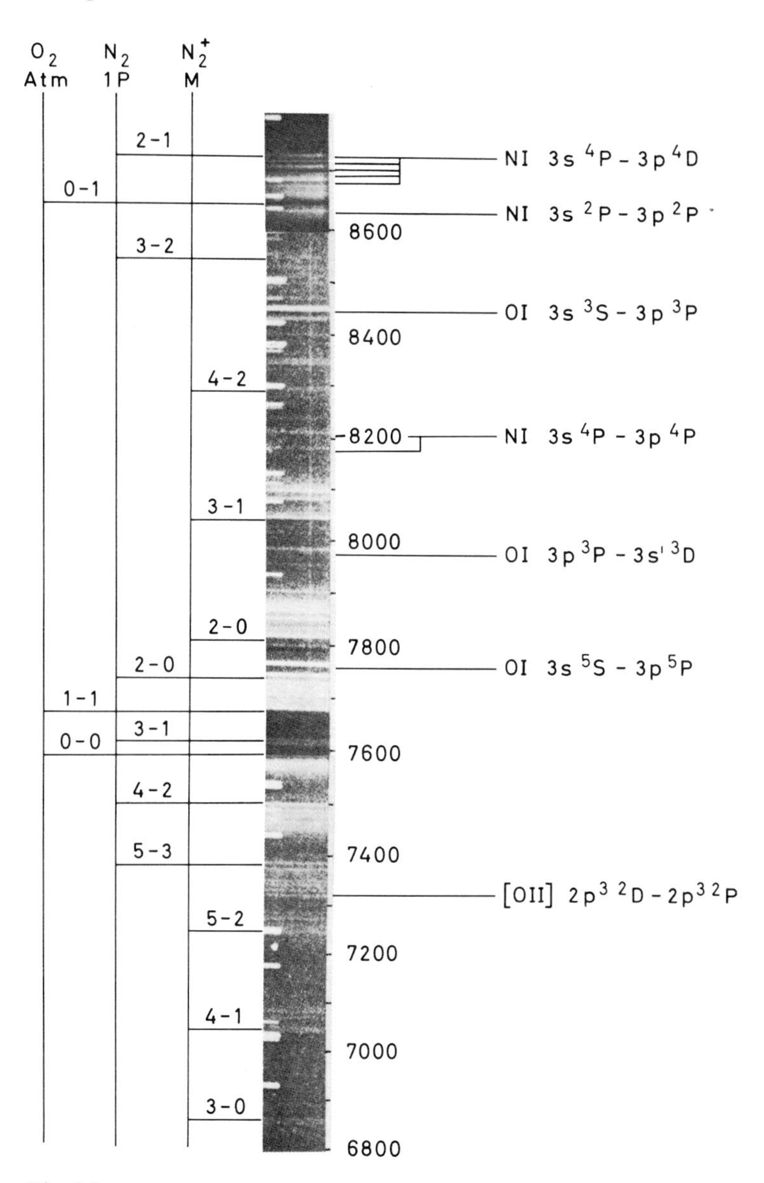

O2 Atm
N2 1P
N2+ M
2-1
0-1
3-2
4-2
3-1
2-0
2-0
1-1
3-1
0-0
4-2
5-3
5-2
4-1
3-0
NI 3s 4P - 3p 4D
NI 3s 2P - 3p 2P
8600
OI 3s 3S - 3p 3P
8400
8200
NI 3s 4P - 3p 4P
8000
OI 3p 3P - 3s' 3D
7800
OI 3s 5S - 3p 5P
7600
7400
[OII] 2p3 2D - 2p3 2P
7200
7000
6800

Fig. 4.6

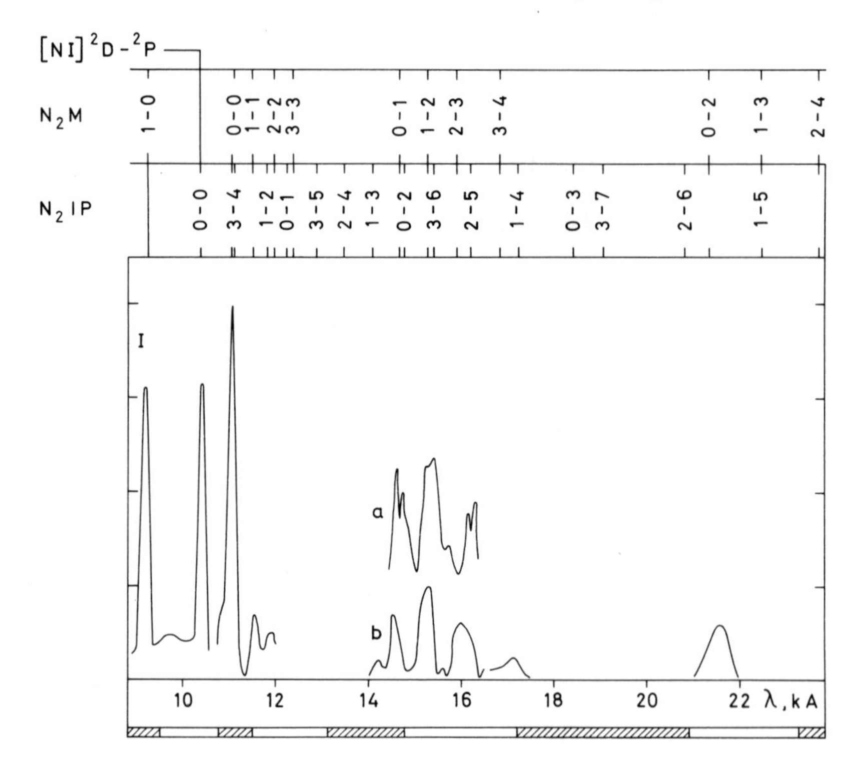

Fig. 4.7 Infrared spectra of aurora, from data collected by Chamberlain (1961) and Vallance Jones (1964). Wavelength in kÅ (1000 Å). Predicted N_2 bands are shown on the top. The (2—2), (3—3), (4—4), and (5—5) bands of $N_2$1P lie on the shortward side of the (0—0) band, between 9.9 and 9.2 kÅ. Hatched areas along bottom line indicate wavelength regions where absorption due to atmospheric water vapour seriously distorts intensity distributions observed from the ground. Spectra a and b give apparent different intensity ratio between first positive and Meinel bands. The feature at 10.4 kÅ is mainly due to the [NI] line, and that at 11.1 kÅ to the (0—0) Meinel band (cf. Harrison 1969)

The identifications shown in Figs. 4.1 to 4.6 are all considered certain. A few lines and bands which do not appear in these particular spectra are included in the identifying text and their positions indicated. The highest resolution has been obtained in the region which is accessible through ground observations using high resolution spectrographs and conventional photographic techniques, i.e. the region from about 3300 Å to about 9000 Å. More detailed information on this region will be found in the book by Chamberlain and in a number of papers by Vegard and co-workers (cf. Chamberlain 1961 and references therein). A report by

Remy, Arpigny and Rosen (1960) gives a great number of lines and identifications, although many of the weakest features claimed to be present in the auroral spectrum must be considered with skepticism. Exact wavelengths are not included in the figures. These are given by Chamberlain (1961) or may be found in suitable reference books, such as that by Wiese, Smith and Glennon (1966) for atomic lines. A bibliography on atomic transition probabilities has recently been published by Miles and Wiese (1970).

4.1.3 Forbidden Atomic Emissions

Emissions have been observed from all excited metastable states of the ground configurations of neutral and singly ionized atomic nitrogen and oxygen. The emissions which appear most regularly are those from the $^1D-{}^1S$ (5577 Å) and the $^3P-{}^1D$ (6300 and 6364 Å) transitions in atomic oxygen and those from the $^2D-{}^2P$ (10395 and 10404 Å) and $^4S-{}^2P$ (3466 Å) transitions in atomic nitrogen.

These so-called forbidden emissions are of particular interest for several reasons: i) Some of them (in particular those at 5577 Å and

Table 4.1 *Forbidden atomic lines identified in auroral spectra*

Atom	Multiplet	Exc. pot. eV	J	Lifetime of excited state seconds	λ(Å)
N	$2p^3\,{}^2D^\circ-2p^3\,{}^2P^\circ$	3.56	$\frac{5}{2}-$	12	10.395.4
			$\frac{3}{2}-$		10.404.1
N	$2p^3\,{}^4S^\circ-2p^3\,{}^2P^\circ$	3.56	$\frac{3}{2}-$		3466.4
N	$2p^3\,{}^4S^\circ-2p^3\,{}^2D^\circ$	2.37	$\frac{3}{2}-\frac{3}{2}$	9×10^4 (26h)	5198.5
			$\frac{3}{2}-\frac{5}{2}$		5200.7
N^+	$2p^2\,{}^1D-2p^2\,{}^1S$	4.04	$2-0$	0.9	5754.8
N^+	$2p^2\,{}^3P-2p^2\,{}^1D$	1.89	$2-2$	250	6583.6
O	$2p^4\,{}^1D-2p^4\,{}^1S$	4.17	$2-0$	0.74	5577.35
O	$2p^4\,{}^3P-2p^4\,{}^1S$	4.17	$1-0$		2972.3
O	$2p^4\,{}^3P-2p^4\,{}^1D$	1.96	$2-2$	110	6300.23
			$1-2$		6363.88
O	$2p^4\,{}^3P-2p^3\,3s\,{}^5S$		$2-2$	6.10^{-4}	1355.61
			$1-2$		1358.52
O^+	$2p^3\,{}^2D^\circ-2p^3\,{}^2P^\circ$	5.00	$\frac{5}{2}-\frac{1}{2}$	5	7318.6
			$\frac{5}{2}-\frac{3}{2}$		7319.4
			$\frac{3}{2}-\frac{1}{2}$		7329.9
			$\frac{3}{2}-\frac{3}{2}$		7330.7
O^+	$2p^3\,{}^4S^\circ-2p^3\,{}^2D^\circ$	3.31	$\frac{3}{2}-\frac{3}{2}$	1.3×10^4 (3.6h)	3726.16
			$\frac{3}{2}-\frac{5}{2}$		3728.91

References: cf. Wiese, Smith and Glennon 1966 and references cited in Table 5.3.

6300/64 Å) are reasonably strong and easy to measure. ii) Theoretical data are available to relate the excitation rates to the impact of particles. iii) The metastable atoms are subject to important deactivating collisions, which transfer energy to other atoms and molecules.

The observed lines are summarized in Table 4.1, and are further discussed in Chapt. 5.

4.1.4 Permitted Atomic Emissions

Table 4.2 shows the transition arrays* observed in the auroral spectrum together with the approximate excitation potentials and the wavelengths or wavelength regions of the optical emission due to each particular transition array. As would be expected, most of the emissions from atomic and singly ionized atomic nitrogen and oxygen are from lower, singly excited configurations. In the case of O and O^+ also transitions from doubly excited configurations occur (with the parent configuration of three or two 2p atoms in an excited state, to which an outer electron is added, also in an excited state).

Lines from higher excited states or other doubly excited states may occur, but if they do they are too weak to be observed with sufficient intensity above noise level and background of continuous light (starlight and stray light) or other emissions. This is what would be expected from theoretical reasoning.

There is also a tentative identification of lines from doubly ionized oxygen and nitrogen. Wallace (1959) claims to have observed two lines, at 3760 and 4379 Å, which may be identified with transitions in O^{++} and N^{++} respectively. The latter of these lines has also been observed by Vegard and associates (cf. Vegard and Kvifte 1951). Undoubtedly double ionization and excitation of atomic oxygen and nitrogen may occur in aurora, but at a very low rate.

Hydrogen lines are present in the particular type of aurora caused by protons, as was discussed in Chapt. 3. Helium lines are occasionally observed; the multiplet at 10830 Å ($^3S-{}^3P$) has been observed in sunlit aurora and that at 5876 Å ($^3P-{}^3D$) on rare occasions in night-time aurora. These emissions are discussed in Sect. 5.5, where references are included.

The sodium doublet at 5890/96 Å was observed in aurora by Vegard and associates (cf. Vegard *et al.* 1955) and by Hunten (1955). This doublet occurs as a regular emission in the nightglow, and the identification as a true auroral feature was ambiguous until Derblom (1964) made con-

* A transition array includes all transitions between energy levels arising from one electron configuration to energy levels arising from another.

clusive observations. Undoubtedly this emission is occasionally enhanced in aurora (cf. Sect. 5.1.2).

4.1.5 Molecular Emissions

The molecular band systems identified in the auroral spectrum are shown in Figs. 4.1 to 4.7 and the transitions are all listed in Table 4.3.

The strong dominating band systems are the permitted ones. The visible part of the spectrum is dominated by the first and second positive N_2 bands and the first negative N_2^+ bands, while the Meinel N_2^+ bands and the Lyman-Birge-Hopfield bands are strong in the red to near-

Table 4.2 *Permitted atomic transitions identified in auroral spectra*

Atom	Parent term	Transition	Exc. pot [1] (eV)	Wavelength or region
N [2]	$2s^2 2p^2(^3P)$	2p−3s	10.5	U.V.
		3s−3p	11.8	I.R.
		3s−4p	13.1	Blue
N^+ [2]	$2s^2 2p(^2P)$	3s−3p	21	Violet-Red
		3p−3d	23	Blue-Green
		3d−4f	26	Violet
O [2]	$2s^2 2p^3(^4S)$	2p−3s	9.5	U.V.
		3s−3p	10.8	Red-I.R.
		3s−4p	12.3	Violet
		3s−5p	12.8	U.V.
		3p−4d	12.8	Red
		3p−5s	12.6	Red
		3p−4d	12.7	Red
		3p−5d	13.1	Green
		3p−6s	13.0	Green
		3p−6d	13.2	Blue
	$2s^2 2p^3(^2D)$	$(^4S)3p-(^2D)3s'$	12.5	Blue
O^+ [2]	$2s^2 2p^2(^3P)$	3s−3p	26	Blue-Violet
		3p−3d	29	Blue-Violet
	$2s^2 2p^2(^1D)$	3s′−3p′	28.5	Blue-Violet
		3p′−3d′	31	Violet
H [3]		$2-3\ H\alpha$	12.0	6563 Å
		$2-4\ H\beta$	12.7	4861 Å
		$2-5\ H\gamma$	13.0	4340 Å
He [4]	$1s(^2S)$	$2s\,^3S-2p\,^3P$	20.9	10830 Å
		$2p\,^3P-3d\,^3D$	23.1	5876 Å
Na [4]	$2p^6\,^1S$	$3s\,^2S-3p\,^2P$	2.1	5890/96 Å

[1] Above ground state of the neutral atom.
[2] For references see Chamberlain (1961).
[3] Cf. Chapt. 3.
[4] For references cf. the text.

Table 4.3 *Molecular transitions identified in the auroral spectrum*

Molecule	Transition	Exc. pot[1)] eV	Wavelength range	Band system
N_2	$A\,^3\Sigma_u^+ \rightarrow X\,^1\Sigma_g^+$	6.2	Blue-U.V.	Vegard-Kaplan (VK)
	$B\,^3\Pi_g \rightarrow A\,^3\Sigma_u^+$	7.4	Red-I.R.	First positive (1 P)
	$C\,^3\Pi_u \rightarrow B\,^3\Pi_g$	11.1	Blue-U.V.	Second positive (2 P)
	$a\,^1\Pi_g \rightarrow X\,^1\Sigma_g^+$	8.6	U.V.	Lyman-Birge-Hopfield (LBH)
N_2^+	$A\,^2\Pi_u \rightarrow X\,^2\Sigma_g^+$	16.7	Red-I.R.	Meinel (M)
	$B\,^2\Sigma_u^+ \rightarrow X\,^2\Sigma_g^+$	18.7	Blue-U.V.	First negative (1 N)
O_2	$b\,^1\Sigma_g^+ \rightarrow X\,^3\Sigma_g^-$	1.6	Red-I.R.	Atmospheric (Atm)
	$a\,^1\Delta_g \rightarrow X\,^3\Sigma_g^-$	1.0	Infrared	Infrared atmospheric[2)] (IRAtm)
O_2^+	$B\,^4\Sigma_g^- \rightarrow A\,^4\Pi_u$	18.0	Green-Red	First negative (1 N)

[1)] Above ground state of neutral molecule.
[2)] The 0—0 band at 12700 Å has been observed from rockets (Megill [*et al.*, 1970) and from balloons and aircraft (Llewellyn *et al.* 1969, cf. also Megill *et al.* 1970 and Noxon 1970).

infrared and in the ultraviolet regions respectively. The latter band system is permitted as magnetic dipole transitions only, but its transition probability is fairly high, about $10^4\ s^{-1}$ (Wilkinson and Mulliken 1959), and collisional deactivation is not probable. The Vegard-Kaplan N_2 bands and the atmospheric O_2 bands, which are due to forbidden transitions, are usually present, but with varying intensity compared to the other, main emissions. The forbidden infrared atmospheric O_2 band system was earlier tentatively included in the list of auroral emissions (Chamberlain 1961), but it is not likely to be present in spectra recorded from the ground (Vallance Jones 1964). However, the (0—0) band at 12700 Å, which is heavily absorbed in the atmosphere (cf. Evans *et al.* 1970), has recently been observed from rockets with great intensity (Megill *et al.* 1970) and from balloons and aircraft (Llewellyn *et al.* 1969, cf. Noxon 1970 also Megill *et al.* 1970).

The first negative O_2^+ bands are weak, but persistent, and most easily observed in low-lying auroras.

From time to time extremely weak spectral features have tentatively been identified with other band systems. Remy *et al.* (1960) made a systematic search for such features, and found that the presence of some other band systems, such as the Goldstein-Kaplan N_2 bands, the second negative N_2^+ bands, the β-system of NO and various systems of NH, as well as enhancements of OH bands, could not be excluded. However, none of these band systems should be considered as conclusively identified, and they are so weak that little, if any, physical information can be drawn from their presence or intensity. A search for enhancement of the nightglow OH emission in the infrared during aurora, made by Harrison (1970), was negative.

4.1.6 The Intensity Distribution in the Spectrum

The intensities of the various lines and bands in the auroral spectrum are of interest not only for identification purposes, but for the study of basic processes as well. As our quantitative knowledge about the various excitation and deactivation processes increases from theoretical and laboratory studies, spectral intensities become increasingly informative. Unfortunately, as pointed out in Sect. 4.1.1, intensity measurements are still far from satisfactory. However, for the stronger emissions valuable data have now begun to emerge.

Prior to about 1953, photographic spectra were the only available means of obtaining intensity distributions in the auroral spectrum. Vegard and his associates (cf. e.g. Vegard and Kvifte 1951, Vegard 1956) made intensity measurements of a great number of auroral lines and bands. One must assume that these intensity measurements were subject to great errors, as was often the case with photographic measurements. The differences between the auroral and the comparison spectra (line spectrum with long exposure time versus continuous spectra with short exposure time) and uncertainties in the continuous standard lamp must be considered as important sources of error. Comparison with later photometric data confirms this. Nevertheless, such measurements were very valuable in the identification process.

Petrie and Small (1952a, b, 1953) have also made extensive intensity measurements in the auroral spectrum. Again the intensities cannot be considered very accurate.

The introduction of photoelectric detectors (photomultipliers) revolutionized the intensity measurements of auroral spectra. Hunten (1953) introduced the scanning photoelectric photometer in auroral research and showed the possibility of obtaining good intensity measurements of auroral spectra. Since then, photoelectric detectors have dominated the intensity measurements (cf. e.g. Hunten *et al.* 1967) and extremely important extensions of the accessible region have been made into the infrared (cf. Vallance Jones 1964, Huppi 1966) and ultraviolet (cf. Fastie 1967). Even with these devices, however, considerable uncertainties are involved; in particular the standard sources for calibration induce errors. Because of the difficulty in obtaining high accuracy, there is still a serious lack of data, in particular for the weaker emissions.

Table 4.4 gives approximate intensities for some important lines and bands, based on data from the references given in the footnotes.

For a discussion of the excitation of the auroral spectrum, the detailed intensity distribution within each molecular band system is important. Some such measurements have been made and will be discussed in detail in Sect. 5.3.2 and in Chapt. 7.

Table 4.4 *The intensity distribution in ordinary auroras*

Atom/ Molecule	Emission	Intensity (rel.)	Ref.	Theoret. int. cf. Table 5.1[1)]
N	$^4S-{}^4P$ λ 1200	3–5	1	
	$^2D-{}^2P$ λ 1493	3	1	
	$^2D-{}^2P$ λ 10400	100	2	
	$^4S-{}^2P$ λ 3466	5	2	
	$^4S-{}^2D$ λ 5199	1	2	
O	$^1D-{}^1S$ λ 5577	100	3	50
	$^3P-{}^1D$ λ 6300/64	20–100	4	450
	$^3P-{}^5S$ λ 1356	~5	1	45
	$^3P-{}^3S$ λ 1304	50–100	1	90
	$^5S-{}^5P$ λ 7773	5–12 } Ratio 0.5	5	2
	$^3S-{}^3P$ λ 8446	10–25 }	5	40
N_2	B–A First pos. b.	500–2400	6	600
	C–B Second pos. b.	50–400	7	100
	a–X L-B-H bands	50–100	8	650
	A–X Vegard-Kaplan b.	~100	9	1300
N_2^+	A–X Meinel b.	700–2000	10	350
	B–X First neg. b.	80–100	11	80
O_2	a–X Infrared atmospheric (0—0) b.	10^4–10^5	12	2300
	b–X Atmospheric (0—0) b.	~200	12	6
O_2^+	B–A First neg. b.	40–100	13	70

[1)] Normalized to the first negative bands.

Table 4.4 References

1. Uncertain. Barth 1967, Fastie 1967, Miller, Fastie and Isler 1968, Barth and Schaffner 1970, Hicks and Chubb 1970. Opal, Moos and Fastie 1970.
2. Cf. Chamberlain 1961.
3. All other intensities are normalized to this value for λ 5577. Note that $I(5577)/I(3914)=2$ has been used in this table for further normalization, cf. ref. 11.
4. Strongly varying. Cf. Chamberlain 1961, Vallance Jones 1969 and Sect. 4.2.3 and Chapt. 5.
5. Omholt 1957, also Percival and Seaton 1956.
6. Vallance Jones 1964, 1969, Shemansky and Vallance Jones 1968.
7. cf. Vallance Jones 1969, Broadfoot 1967.

8. Uncertain. Miller, Fastie and Isler 1968, Opal, Moos and Fastie 1970.
9. Very uncertain, cf. Chamberlain 1961.
10. Vallance Jones 1964, 1969, Shemansky and Vallance Jones 1968.
11. Adopting 2.0 as the most common value for $I(5577)/I(3914)$, cf. Sect. 4.2.2.
12. Megill, Despain, Baker and Baker 1970, Noxon 1970.
13. Shemansky and Vallance Jones 1968, also Vallance Jones 1969.

4.2 Local Variations in the Spectrum

4.2.1 Introduction

By local variations in the spectrum we mean variations that are related to the height of the emission or its relative position within an auroral form. The most pronounced variations in the auroral spectrum seem to be associated with variations in the height of the aurora. This is in agreement with current knowledge about the basic excitation processes. Electrons or protons of a given energy deposit their energy within a restricted altitude range in the atmosphere. Hence the excitation at a particular height is largely performed by electrons or protons within a restricted energy interval. The kind of particle and its specific energy, together with the atmospheric density and composition, determine the spectral characteristics.

Spectral differences which have been noted between different types of aurora are probably also due to height variations. Type A aurora and red arcs, which are red auroras with strong red [OI] lines, are both thought to occur at high altitudes, whereas Type B aurora, which is characterized by a strong, red lower border, occurs very low in the atmosphere (70—90 km).

It seems most probable, therefore, that the different spectral types of aurora are only the extremes of a spectral characteristic which changes continuously with height. Hence, we shall leave out any explicit discussion of particular types of aurora, but rather describe height variations of the spectrum in very general terms.

Although important spectral variations with height and horizontal direction have been noted and observed, good quantitative measurements have been difficult to obtain, for several reasons. One is the complicated geometry of the aurora together with the fact that an observer on the ground automatically integrates the optical emissions along the line of sight of his instrument. Triangulation of auroral forms, confined to particular spectral lines, has been difficult, and not until recently has this been done with some success.

Another difficulty is the comparison of emissions in different spectral regions. Quantitative measurements of spectral intensities is a difficult subject, and even relative measurements have been seriously affected by calibration errors when the wavelength difference is great.

Finally, atmospheric scattering and absorption distort the intensity distribution to a considerable degree, in particular at shorter wavelengths, where Rayleigh scattering is serious, and in the infra-red, where absorption is dominating. Corrections for scattering and for absorption by air constituents which vary in amount are always difficult. The standard extinction tables used by astronomers are valid for point sources only. In the case of an extended light source, the light which is scattered out of the beam is partly compensated by light scattered into the beam, but originating from other parts of the source. With an aurora of even intensity covering the whole sky, the apparent attenuation due to scattering is reduced to about one half, since altogether about half of the scattered light escapes upwards while the other is proceeding downwards. With varying amounts of water vapour, dust or other, often man-made, pollution the scattering and absorption may vary considerably from time to time.

For these reasons, much of the data presented here are rather qualitative; better quantitative measurements are urgently needed.

4.2.2 The $I(5577)/I(1\text{ neg})$ Ratio

The ratio between the intensities of the green [OI] line $^1D-{}^1S$ at 5577 Å and the first negative N_2^+ bands is of considerable importance. The two emissions have widely different excitation potentials. For O^1S it is 4 eV if excited from O^3P, 9 eV if excited through dissociation of O_2 (cf. Sect. 5.1.1 and 5.3.1) and for the first negative bands it is about 19 eV. The excitation ratio should therefore be sensitive to changes in excitation conditions. The strongest and most important first negative bands are the (0—0), (0—1) and (0—2) bands at 3914, 4278 and 4709 Å respectively.

Knowledge about variations in the $I(5577)/I(1\text{ neg})$ ratio is also important because each emission is widely used as an indicator of auroral activity and to monitor the general auroral intensity, being among the strongest and most easily measured emissions. If their relative intensity varies in any systematic or erratic way, it is imperative to know this before comparing different sets of observations.

It has been generally accepted that the $I(5577)/I(1\text{ neg})$ ratio is reasonably constant (within $\pm 50\%$) in ordinary auroras. Although absolute calibration difficulties induce greater differences between results of different authors, each individual set of observations confirms that this usually is the upper limit of variation (Rees 1959, Omholt 1959,

Malville 1959, O'Brien and Taylor 1964, Henriksen 1969, Mende *et al.* 1970).

There do not seem to be large height variations within an auroral form, at least up to 200 km for auroras of moderate intensity. This is borne out from ground-based as well as rocket observations (Romick and Belon 1967, Baker 1967, Cummings *et al.* 1966, Ulwick 1967, Maseide, private communication, Parkinson *et al.* 1970).

In very low auroras the $I(5577)/I(1\text{ neg})$ ratio is lower. In auroras associated with PCA events a ratio of 0.2 or less is found (cf. Sect. 8.6) and the emission is probably located between 30 and 80 km altitude. Similarly, type B aurora, which has a lower border at 80 km or so, shows a significantly lower intensity ratio (Evans and Vallance Jones 1965).

There is, on the average, apparently no great systematic variation in this ratio with form and intensity, although it may be inferred from

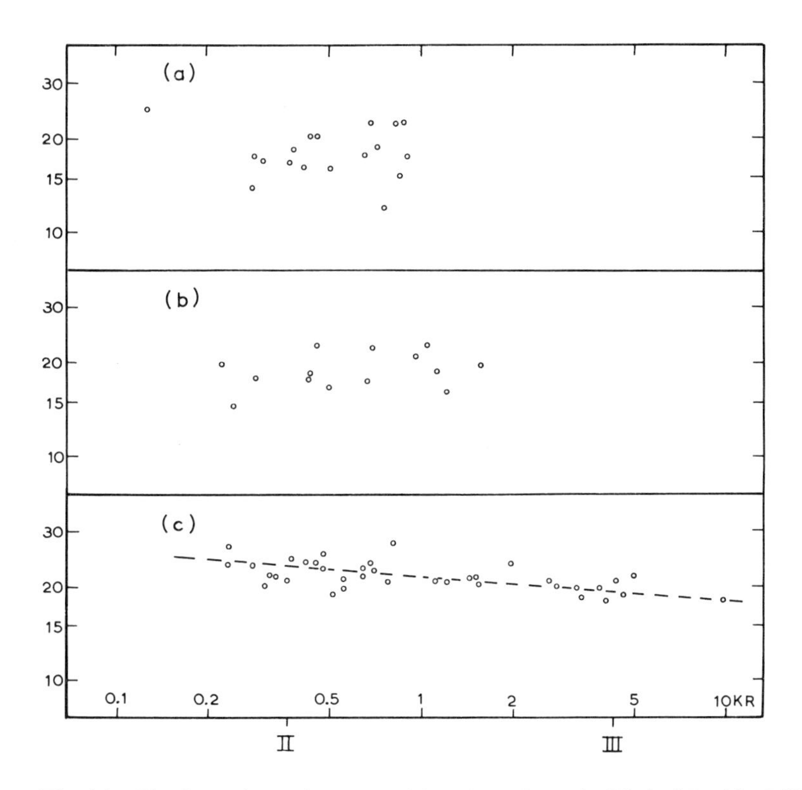

Fig. 4.8 The intensity ratio $I(5577)/I(4709)$ against $I(4709)$ in kR. (a): Diffuse and patchy aurora, glow. (b): Upper and middle part of rays. (c): All other types of aurora

data published by Kawajiri *et al.* (1965) that there are sometimes systematic variations with intensity, since the two emissions apparently relate differently to the associated sporadic E-layer ionization. The data shown in Fig. 4.8 indicate a slight systematic decrease in $I(5577)/I(4709)$ with increasing absolute intensity, whereas O'Brien and Taylor's (1964) and Henriksen's (1969) data give no indication of such an effect.

Systematic, large spatial variations in quiet arcs have been reported observed with scanning photometers (Romick and Belon 1967), but it is doubtful whether this is due to anything but insufficient correction for atmospheric scattering (Mende, Eather and Evans 1970). Similar variations have been reported from observations taken sideways from rockets (Murcray 1967, Baker 1967, Monfils 1968, Duysinx and Monfils 1970). The interpretation of these observations is, however, rather ambiguous, particularly because the photometers sweep rapidly around. If the auroral intensity varies rapidly, then large transient variations in the $I(5577)/I(3914)$ ratio may occur at any one spot, due to the delay time of emission of the forbidden green line (cf. Sect. 5.4). However, it seems possible that some systematic variations occasionally occur, the $\lambda\,5577$ emission being more restricted in horizontal direction than the first negative N_2^+ bands.

From ground observations too, variations as large as a factor of 3 have been noted occasionally in the $I(5577)/I(4278)$ ratio, the green line being much more enhanced in the center of the form than the N_2^+ bands (Brekke and Omholt 1968).

We may conclude that at least when integrated over a considerable auroral volume (i.e. a large field of view) the $I(5577)/I(1\text{ neg})$ ratio is fairly constant. From the data given by Rees (1959), Baker (1967), Monfils (1968) and Omholt (1959) a value of 2.0 is derived for the $I(5577)/I(3914)$ ratio (with units in kR and not energy). The ratio 1.0:0.34:0.075 is then adopted for the intensities of the $\lambda\,3914$, $\lambda\,4278$ and $\lambda\,4709$ bands (cf. Sect. 2.2). The variation with height is small except for low altitudes, whereas there are indications that spatial variations occasionally occur in the detailed picture.

4.2.3 Other Forbidden Emissions

In general, forbidden lines and bands which are emitted from metastable atoms and molecules show an intensity-versus-altitude profile which is different from permitted emissions. Collisional deactivation (cf. Sect. 5.2) quenches these emissions at lower altitudes, and therefore the intensities relative to other emissions decrease strongly with decreasing height.

So far the best-studied emission from long-lived states is the red [OI] multiplet at 6300 and 6364 Å ($^3P-{}^1D$). It has been known for a long time that high-altitude auroras are associated with relatively strong red oxygen lines (cf. Chamberlain 1961, Harang 1958). Quantitative measurements of this effect, however, proved difficult, and later erroneous. This is due to the large differences in the spatial distribution of the red lines and the permitted emissions. Not only does quenching occur at low heights, but also the excited atoms diffuse outward from the auroral form during their considerable lifetime (110 s if undisturbed). At great heights even different excitation mechanisms occur (cf. Sects. 5.1.2 and 5.1.3). These effects were clearly demonstrated in the auroral arc triangulated with photometer techniques by Romick and Belon (1967).

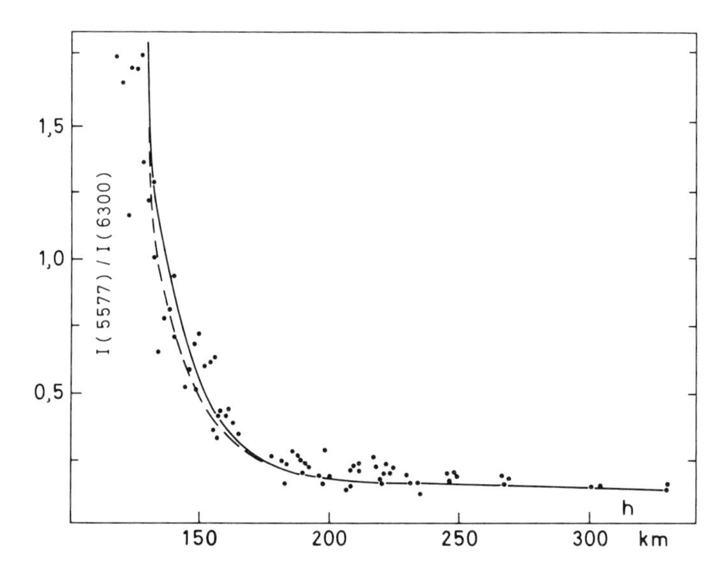

Fig. 4.9 $I(5577)/I(6300)$ versus height, from rocket data by Murcray (1969). Solid and dashed curves are computed ones for deactivation by respectively air and nitrogen only. Deactivation coefficient 10^{-12} cm^3 s^{-1} ($\alpha = 10^{-10}$ in Murcray's paper is erroneously taken to be the deactivation coefficient, while it also includes the lifetime against radiation of the excited state)

The λ 6300 emission showed a peak between 220 and 300 km, depending on the assumed horizontal distribution, whereas the [OI] λ 5577 line and the N_2^+ first negative bands had their maxima at about 120 km. The red line intensity distribution was also much wider in horizontal extent, more than 10 times that of the green line (a half width of about 50 km compared to slightly less than 4 km). Romick and Belon concluded that collisional deactivation of the red lines is essentially complete below 150 km, and significant up to a height of 240 km. Also, their measurements indicate an almost infinite $I(6300)/I(5577)$ ratio at the top of the aurora, in agreement with what has occasionally been found by others (cf. Chamberlain 1961). Also the high red arcs, which occupy the F-region of the ionosphere between 300 and 700 km, are monochromatic in the λ 6300/63 line (cf. e. g. Elvey 1965). Although triangulation was performed by Romick and Belon (1967) on a single aurora only, probably being subject to considerable errors, the effects are well demonstrated.

Murcray (1969) has measured the $I(5577)/I(6300)$ ratio from rocket photometers, and found a height variation as shown in Fig. 4.9. His data indicate that the intensity ratio in question is approximately constant between 300 and 200 km, and that quenching is serious, but far from complete, at 150 km. 90% quenching of the red lines seems to occur at about 130 km. The quenching rate coefficient of about 10^{-12} $cm^3\ s^{-1}$ which may be derived from Murcray's data is lower than inferred from other data (cf. Sect. 5.2). Jorjgo (1960) found that the $I(6300)/I(5577)$ ratio decreased with increasing intensity of the aurora.

Only qualitative information exists on the height variation of forbidden lines and bands other than [OI]. It is clear, however, that the height variation is similar to that of the red lines. Thus, the [OII] multiplet ($^2D-^2P$) (centered at 7319 and 7330 Å) was first detected in spectra from the upper part of auroras (Omholt 1957). Similarly, the [NII] line at 6584 Å ($^3P-^1D$) has been observed in type A auroras (Belon and Clark 1959) and correlated with narrow hydrogen lines (Harang and Pettersen 1967).

The O_2 atmospheric bands ($b\,^1\Sigma_g^+ \rightarrow X\,^3\Sigma_g^-$) were found to increase strongly relative to the first positive N_2 bands with increasing height (indicated by elevation angle above horizon and the relative intensity of the red oxygen line), despite the fact that the relative content of O_2 decreases upwards (cf. Fig. 4.10).

4.2.4 Permitted Atomic Lines

There are observations which indicate that also permitted atomic lines, and in particular lines from atomic ions, are relatively stronger in high auroras than in lower ones. For example, visual inspection of the spectra illustrating Chamberlain's (1961) book does give this impression.

Fig. 4.10 shows the relative intensity of some lines and bands correlated with that of the [OI] λ 6300 line. The intensities (RI) are relative to that of the (2–0) and (4–0) first positive bands. Points from the same aurora correspond, from left to right, to the lower, middle and upper parts of the aurora (Omholt 1957), respectively.

This figure shows that in particular the λ 7774 ($3s\,^5S-3p\,^5P$) and λ 8446 ($3s\,^3S-3p\,^3P$) OI lines increase strongly in relative intensity upwards. Incidentally, this agrees qualitatively with Stolarski's (1968) computations on electron impact excitation of the lower of the states in question here. As for the intensity ratio between the two, however, agreement with theoretical data is rather poor (cf. Sect. 5.7).

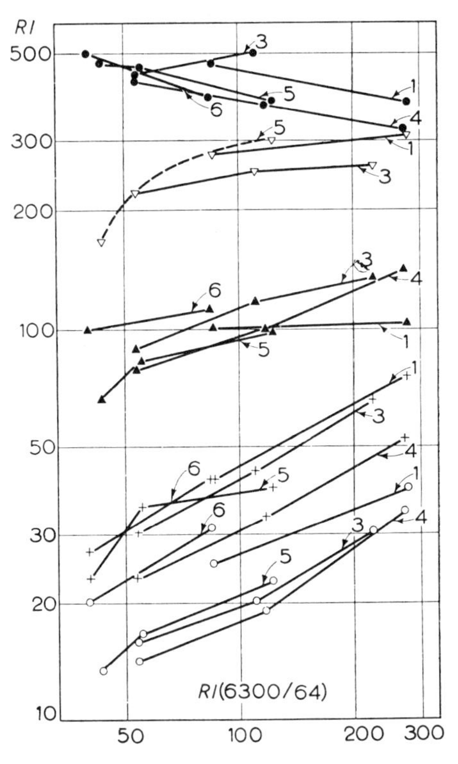

Fig. 4.10 Relative intensities (RI) for some spectral lines and bands against RI (6300/64) (for definition of RI see the text). Lines are drawn between points from the same aurora. ● N_2^+ Meinel (2—0) + (3—1); $RI\times 4$. ▲ O_2 Atmospheric (1—1); $RI\times 5$. ▽ [OI] λ 5577. ○ OI λ 7774. + OI λ 8446

The same set of measurements (Omholt 1957) indicates little variation with height of the relative intensity of the OI λ 7990 ($3p\,^3P-3s'\,^3D$) and the NI λ 8186 and λ 8216 ($3s\,^4P-3p\,^4P$) multiplets, but the data were less reliable for these lines. Miller, Fastie and Isler (1968) measured the height variations of the [OI] line at 1356 Å ($^3P-{}^5S$), the NI line at 1200 Å ($^4S-{}^4P$), and the Lyman-Birge-Hopfield N_2 bands. They find that all these emissions vary with height like the first negative

N_2^+ bands. The O I line at 1304 Å ($^3P - {^3S}$) shows a particular height distribution because of imprisoning, since it is a permitted resonance radiation.

4.2.5 Permitted Molecular Bands

The data shown in Fig. 4.10 indicate that the Meinel N_2^+ bands decrease relative to the first negative N_2^+ bands with increasing altitude, provided the $I(5577)/I(1\,\text{neg})$ ratio is constant (this seems to be so except at the lowest height covered in the figure, cf. Sect. 4.2.2).

The red lower border of the type B aurora, which probably occurs at heights between 70 and 90 km, was earlier attributed to an enhancement of the N_2 first positive system. Spectrographic measurements indicated an enhancement by a factor of about 1.5 over the first negative N_2^+ bands (Vegard and Tønsberg 1937, Vegard 1940, Malville 1959, Evans and Vallance Jones 1965). Shemansky and Vallance Jones (1968), however, found no enhancement of the first positive bands over the Meinel N_2^+ bands. On the contrary, the ratio $I(\text{Meinel})/I(1\,\text{pos.})$ was found to be higher in type B than in ordinary auroras. This is qualitatively in agreement with the data in Fig. 4.10. Shemansky and Vallance Jones (1968) concluded, on the basis of this, that there occurs no enhancement of the first positive bands relative to the first negative bands, assuming that the Meinel bands are well correlated with first negative ones. A detailed discussion of the difficulties inherent in all quantitative measurements of the auroral spectrum provided the explanation for the apparently erroneous earlier results. However, the data in Fig. 4.10 may indicate that $I(\text{Meinel})/I(1\,\text{neg.})$ is not constant with height, and its possible variation is such that it may provide an explanation for the disagreement between Shemansky and Vallance Jones' (1968) conclusion and that of earlier investigators, i.e. there might be an enhancement of both the first positive and the Meinel bands in the lower part of type B aurora.

Opal, Moos and Fastie (1970) found with rocket measurements that the intensity of the Lyman-Birge-Hopfield N_2 bands fell off more rapidly than the first negative N_2^+ bands towards the bottom of an aurora, the ratio $I(L-B-H)/I(1\,\text{neg.})$ decreasing by a factor of 10 from 110 to 100 km altitude.

It may be worthwhile to keep in mind that none of the measurements cited are very accurate, and that Stolarski's (1968) computations do not give any large height variations in the intensity distribution among the N_2 and N_2^+ band systems.

Enhancement of the O_2^+ first negative bands in type B aurora has been found by Hunten (1955) and Shemansky and Vallance Jones (1968) and was also indicated by photometer measurements of Evans and

Vallance Jones (1965). Relative to the first negative bands there is an enhancement by a factor of about 2—3. The explanation of this enhancement presents no particular difficulty, type B auroras occurring typically at altitudes below the O_2-O transition region. There is no evidence that excitation mechanisms other than simultaneous ionization and excitation by fast particles and secondaries need be considered.

Shemansky and Vallance Jones (1968) also find a different vibrational development of the N_2 first positive bands in type B auroras and high, weak auroras (cf. Table 4.5, and also Fig. 5.2). The distribution in the high aurora seems largely consistent with electron excitation and a 5% contribution through cascading (through emission of the second positive bands), and there is no apparent explanation for the difference. Large variations in the cascading from the $C^3\Pi_u$ state are unlikely, because this would reveal itself as variation in the ratio $I(1\text{ pos.})/I(2\text{ pos.})$.

Table 4.5 *Relative population* ($N_{v'}$) *for the* $N_2\,B^3\Pi_g$ *state* *

v'	0	1	2	3	4	5	6	7	8	9	10	Source
$N_{v'}$				131	100	47	26	12	8			Hunten: Type-B red
$N_{v'}$				122	100	52	28	14	9			Shemansky: Type-B red
$N_{v'}$				176	100	61	37	21	12			Shemansky: Weak, high level
$N_{v'}$	65	117	143	132	100	67	40	20	12	6	3	Theory, 5% cascade from $C^3\Pi_u$

* (Shemansky and Vallance Jones 1968).

4.3 Latitudinal and Zonal Variations

Satellite measurements by O'Brien and Taylor (1964) of $I(5577)/I(1\text{ neg.})$ between geomagnetic latitudes about 64° and 73° N show no indication of significant variation of this intensity ratio over that latitude range. Eather (1969), measuring from a high-altitude aircraft, found that this intensity ratio showed a significant decrease at latitudes above about 73° N. At 80° N invariant latitude it is only about 2/3 of that in the auroral zone. In numbers, $I(5577)/I(4278)$ was about 3.2 in the auroral zone* and about 2 at 80° invariant latitude.

The most pronounced variation, however, occurs in the intensity of the red [OI] lines at 6300/64 Å relative to the first negative bands. According to Eather's (1969) measurements, the λ 6300 line has an average, and fairly constant, value of about 0.8 that of the λ 4278 N_2^+ bands between 60° and 73° N invariant latitude. At the latter latitude it suddenly increases, reaching a value of 5 at 75° N, and remains high

* This ratio corresponds to about 1 for $I(5577)/I(3914)$ (cf. Sect. 2.2), contrary to a value of about 2 found by other investigators, cf. Sect. 4.2.2.

and slightly fluctuating up to about 83° N, the highest latitude for the measurements. Sandford (1967, 1968) found, from observations in Antarctica, that a similar effect existed there in 1959. Colour films from Scott Base showed that 70% of all discrete visual aurora had an extensive red glow above them. During solar minimum, however, such red line enhancements were rare. That the intensity of λ 6300 was particularly sensitive to the level of geomagnetic disturbances was also noted by Truttse (1968).

The variation in I(6300)/(4278) is in agreement with evidence from height measurements (cf. Sect. 2.7), that high altitude auroras are more common at high geomagnetic latitudes. However, much more work is needed to clarify this point. There is also evidence (Akasofu, private communication) that high red auroras are common on the day side of the auroral oval. A plausible explanation may be that the λ 6300 emission is caused by a separate influx of very low energy electrons.

Other typical latitude variations noted are linked more to particular types of aurora. The polar cap absorption events are associated with auroral glow where $I(5577)/I(3914)$ is 0.2 or less. This effect is directly related to the height of the emission and the collisional deactivation occurring at these heights (cf. Sect. 8.6).

The typical mid-latitude red arcs, on the other hand, show strong enhancement of the red [OI] lines, being in fact observationally monochromatic in the red lines (cf. Sect. 1.2.3). This is also a direct height effect.

Sandford (cf. Sandford 1967) finds that there is a diurnal or zonal variation in $I(3914)/I(5577)$ in the mantle aurora (cf. Sect. 1.2.4), the auroral glow which extends over large areas of the polar cap. At Byrd Station, Antarctica (geom. lat. 69° S) this ratio is between 1 and 2 from 1900 to 0700 geom. time. It then decreases and reaches a minimum between 0.3 and 0.6 slightly after magnetic midday.

There is evidence that the red [OI] lines are also more pronounced at low latitudes, and that a significant intensification of these lines accompanies geomagnetic activity (cf. Barbier 1960, Krassovsky 1964, Truttse 1967). Truttse (1967) showed from IGY data that on magnetically disturbed nights there is a strong correlation between the intensity of the λ 6300 emission and the geomagnetic activity. The ratio between λ 6300 intensity and degree of magnetic disturbance varied systematically during the years 1957—59, and in much the same manner as the thermopause temperature as calculated from the solar radiations at 10.7 and 8 cm. With a high thermopause temperature, a given magnetic disturbance gave a high λ 6300 intensity. This may point in the direction of excitation by thermal electrons, heated through induced currents (cf. Sects. 5.1.3 and 5.1.4).

References

Baker, K. D.: In The Birkeland Symposium on Aurora and Magnetic Storms. (Ed. A. Egeland and J. Holtet) CNRS 1967.
Barbier, D.: Ann. Geophys. **16,** 544 (1960).
Barth, C.A.: In The Birkeland Symposium on Aurora and Magnetic Storms. (Ed. J. Holtet and A. Egeland) CNRS 1967.
— Schaffner, S.: J. Geophys. Res. **75,** 4299 (1970).
Belon, A. E., Clark, K. C.: J. Atmospheric Terrest. Phys. **16,** 220 (1959).
Brekke, A., Omholt. A.: Planetary Space Sci. **16,** 1259 (1968).
Broadfoot, A. L., Hunten, D. M.: Planetary Space Sci. **15,** 1801 (1964).
Chamberlain, J. W.: Physics of the Aurora and Airglow. Academic Press 1961.
Cummings, W. D., La Quey, R. E., O'Brien, B. J., Walt, M.: J. Geophys. Res. **71,** 1399 (1966).
Derblom, H.: J. Atmospheric Terrest. Phys. **26,** 791 (1964).
Duysinx, R., Monfils, A.: J. Geophys. Res. **75,** 2606 (1970).
Eather, R. H.: J. Geophys. Res. **74,** 153 (1969).
Elvey, C. T.: In Auroral Phenomena. (Ed. M. Walt) Stanford University Press. 1965.
Evans, W. F. J., Vallance Jones, A.: Can. J. Phys. **43,** 697 (1965).
— Wood, H. C., Llewellyn, E. J.: Can. J. Phys. **48,** 747 (1970).
Fastie, W. G.: Appl. Opt. **6,** 397 (1967).
Harang, L.: Geofys. Publikasjoner **20,** No. 5 (1958).
Harang, O., Pettersen H.: Planetary Space Sci. **15** 1599 (1967).
Harrison, A. W.: Can. J. Phys. **47,** 599 (1969).
— J. Geophys, Res. **75,** 1330 (1970).
Henriksen, K.: Absolute measurements of the intensity ratio I(5577)/I(4278) in bright aurora. Report The Auroral Observatory, Tromsø 1969.
Hicks, G. T., Chubb, T. A.: J. Geophys. Res. **75,** 1290 (1970).
Hunten, D. M.: Can. J. Phys. **31,** 681 (1953).
— J. Atmospheric. Terrest. Phys. **7,** 141 (1955).
— Rundle, H. N., Shepherd, G. G., Vallance Jones, A.: Appl. Opt. **6,** 1609 (1967).
Huppi, E. R.: Techniques for the measurements of airglow and aurora in the infrared. Sci. Rep. No. 5, Contract No. AF 19 (628—251), Utah State University 1966.
Jorjgo, N. V.: Spectr. Electrophotom and Radar Res. Aurora and Airglow No **2—3,** 45, Moscow 1960.
Kawajiri, N., Wakai, N., Nakamura, J., Nakamura, I., Hasegawa, S.: J. Radio Res. Lab. **12,** 141 (1965).
Krassovski, V. I.: Space Sci. Rev. **3,** 232 (1964).
Llewellyn, E. J., Wood, H. C., Vallance Jones, A.: (Abstract) Trans. Am. Geophys. Union **50,** 271 (1969).
Malville, J. M.: J. Atmospheric Terrest. Phys. **16,** 59 (1959).
Megill, L. R., Despain, A. M., Baker, D. J., Baker, K. D.: J. Geophys. Res. **75,** 4775 (1970).
Mende, S. B., Eather, R. H., Evans, J. E.: Trans. Am. Geophys. Union **51,** 371 (1970).
Miller, R. E., Fastie, N. G., Isler, R. C.: J. Geophys. Res. **73,** 3353 (1968).
Monfils, A.: Space Sci. Rev. **8,** 804 (1968).
Miles, B. M., Wiese, W. L.: Natl. Bureau of Standards. Spec. Publ. 320, 1970.
Murcray, W.: J. Geophys. Res. **72,** 1047 (1967).

Murcray, W.: Planetary Space Sci. **17,** 1429 (1969).
Noxon, J. F.: J. Geophys Res. **75,** 1879 (1970).
O'Brien, B. J., Taylor, H.: J. Geophys. Res. **69,** 45 (1964).
Omholt, A.: J. Atmospheric Terrest. Phys. **10,** 320 (1957).
— Geofys. Publikasjoner **21,** No 1 (1959).
Opal, C. B., Moos, H. W., Fastie, W. G.: J. Geophys. Res. **75,** 788 (1970).
Parkinson, T. D., Zipf, E. C. Jr., Donahue, T. M.: Planetary Space Sci. **18,** 187 (1970).
Percival, I. C., Seaton, M. J.: In The Airglow and the Aurorae. (Ed. E. B. Armstrong and A. Dalgarno) Pergamon Press 1956.
Petrie, W., Small, R.: Astrophys. J. **116,** 433 (1952a).
— — J. Geophys. Res. **57,** 51 (1952b).
— — Can. J. Phys. **31,** 911 (1953).
Rees, M. H.: J. Atmospheric Terrest. Phys. **14,** 325 (1959).
Remy, L., Arpigny, C., Rosen, B.: Identifications in the Spectra of Aurorae. Techn. Note nr. 6. Contract AF 61(052)—24. University of Liège 1960.
Romick, G. J., Belon, A. E.: Planetary Space Sci. **15,** 1695 (1967).
Sandford, B. P.: In Aurora and Airglow. (Ed. B. M. McCormac) Reinhold Publ. Co. 1967.
— J. Atmospheric Terrest. Phys. **30,** 1921 (1968).
Shemansky, D. E., Vallance Jones, A.: Planetary Space Sci. **16,** 1115 (1968).
Stolarski, R. S.: Planetary Space Sci. **16,** 1265 (1968).
Truttse, Yu. L.: Planetary Space Sci. **16,** 140 (1967).
— Planetary Space Sci. **16,** 981 (1968).
Ulwick, J.C.: In Aurora and Airglow. (Ed. B.M. Mc. Cormac) Reinhold Publ. Co. 1967.
Vallance Jones, A.: Mem. Soc. Roy Sci. Liege [5] Ser. **9,** 289 (1964).
— In Atmospheric Emissions. (Ed. B. M. McCormac and A. Omholt) Van Nostrand Reinhold Co. 1969.
Vegard, L.: Terr. Magn. **45,** 5 (1940).
— Geofys. Publikasjoner **19,** No 9 (1956).
— Kvifte, G.: Geofys. Publikasjoner **18,** No 3 (1951).
— — Omholt, A., Larsen, S.: Geofys. Publikasjoner **19,** No 3 (1955).
— Tønsberg, E.: Geofys. Publikasjoner **11,** No 16 (1937).
Wiese, W. L., Smith, M. W., Glennon, B. M.: Atomic Transition Probabilities. Nat. Bur. Standards 1966.
Wallace, L.: J. Atmospheric Terrest. Phys. **17,** 46 (1959).
Wilkinson, P. G., Mullikan, R. S.: J. Chem. Phys. **31,** 674 (1959).

Chapter 5

Physics of the Optical Emissions

5.1 Excitation

Excitation of the auroral spectrum emitted from atmospheric atoms and molecules can be attributed to four classes of process: i) direct excitation by primary particles or secondary electrons, ii) thermal collisions involving ionized or excited atoms and molecules, iii) excitation by heated, ambient electrons, and iv) discharge mechanisms and heating by electric fields.

A useful review of atomic and molecular excitation mechanisms in gases up to 20,000 °K is given by Gilmore, Bauer and McGowan (1969), while Danilov (1970) has given a comprehensive review of the chemistry of the ionosphere.

Excitation of the hydrogen lines, associated with proton impact, was discussed in Chapt. 3.

5.1.1 Fast Particle Impact

The excitation of atmospheric atoms and molecules during auroras is most directly performed by the primary particles and the secondary electrons. Energetic electrons in air produce ions as well as excited atoms and molecules directly, and also secondary electrons which have enough energy to perform secondary ionization and excitation. Even tertiary ionization may occur.

Experimental results (Landolt-Börnstein 1952, Vol. 5, part 1, p. 343) indicate that only about 30% of the total number of the ions and electrons are produced directly by fast primary electrons, while 70% are produced by secondary and higher order processes. This result was supported by theoretical computations by Green and Barth (1965). This would mean that each secondary electron carried away an energy of nearly 90 eV which may be spent on secondary ionization and excitation.

In a later series of papers Green and associates studied the relevant excitation cross-sections in more detail (Green and Dutta 1967, Jusick *et al.* 1967, Stolarski *et al.* 1967, Watson *et al.* 1967), and refined the computations on excitation in aurora (Stolarski and Green 1967). According to these computations a much higher fraction of the ionization processes is due to primary ionization (about 80%) and on the average each secondary electron carries away only 20 eV as kinetic energy. The number of ion pairs produced is about 3 for each 100 eV of energy which is in good agreement with laboratory data (cf. Dalgarno 1962). However, even accepting Stolarski and Green's recent results, it is clear from a comparison of the various relevant cross-sections that excitation of neutrals is performed to a large extent by secondary electrons. This is particularly so for low-lying, metastable states, to the extent these are not excited through a simultaneous ionization or dissociation process.

As far as can be judged from available data, relative excitation and ionization rates are fairly independent of the energy of the primary particle. Both Stolarski and Green's (1967) data and computations on electron impact on hydrogen (cf. Dalgarno 1962) indicate that the energy per ion pair is fairly constant down to energies of about 100 to 200 eV, rising at lower energies. For protons as primary particles, no detailed data appear to exist, but from theoretical considerations it is reasonable to assume that the situation is not drastically different from that of electrons.

For this reason, it is believed that a significant part of the excitation of the auroral spectrum is due to secondary electrons.

Green and Barth (1965) and Stolarski and Green (1967) made detailed studies of the ionization and excitation processes by impact of electrons on N_2 and a mixture of 45% N_2, 45% O and 10% O_2, respectively. The energy loss of the primary particles was calculated from the energy loss function (Stolarski and Green 1967):

$$\begin{aligned} L(E) &= -\frac{1}{n}\frac{\mathrm{d}E}{\mathrm{d}x} \\ &= \sum_s \frac{n_s}{n}\left[\sum_j W_j \sigma_j(E) + \sum_i \int_{I_i}^{E} W S_i(E, W)\,\mathrm{d}W\right]_s \end{aligned} \tag{5.1}$$

In this expression $L(E)$ is the energy loss per molecule within one unit of area along the path, n is the gas density, n_s the density of the sth constituent of the gas, E the particle energy, x the path length, W_j and W the energy losses per collision for excitation and ionization respectively, $\sigma_j(E)$ the excitation cross-section, $S_i(E, W)$ the differential cross-section for ionization with energy loss W, and I_i the ionization potential.

The fractional energy loss to any particular excited state is given by

$$f_j(E) = \frac{n_s}{n} W_j \sigma_j(E)/L(E)\,. \tag{5.2}$$

For ionization,

$$f_i(E) = \frac{n_s}{n} \overline{W}_i \sigma_i(E)/L(E). \tag{5.3}$$

is the fractional energy spent in creating the ion, and

$$f_s(E) = \frac{n_s}{n} (\overline{W}_i - I_i) \sigma_i(E)/L(E) \tag{5.4}$$

is the average energy vested in secondary electrons originating from a particular ionization process. $\overline{W}_i$ is the average value of W.

Stolarski and Green (1967) adopted the cross-sections for all important excited states and for ionization of N_2 and O_2 from the preceding papers by Stolarski *et al.* (1967) and Watson *et al.* (1967), and for O they used mainly theoretical data by Seaton (1956, 1959), which are similar to more recent ones obtained by Smith *et al.* (1967). Excitation of $O(^1D)$ and $O(^1S)$ was considered to arise solely from excitation of atomic oxygen. Parkinson *et al.* (1970) have recently cast doubt on this assumption, which has been implicit in most discussions of the green line λ 5577. On the basis of rocket measurements of λ 5577 and λ 3914 intensities, ion composition and secondary electron spectrum and densities, they find that only dissociation of O_2 with $O(^1S)$ atoms as a frequent product can explain their observations.

Stolarski and Green's results are given in Table 5.1. The emission in the $(0-0)$ first negative N_2^+ band (3914) is normalized to 100. There are relatively large uncertainties in the cross-sections for the various excitation and ionization processes. Therefore the data must be regarded as indicative only. Collisional deactivation will tend to reduce the intensities of forbidden lines and band systems. Also, the atmospheric composition used is most relevant to heights well above that for the main luminosity of an average aurora (cf. Danilov 1970). Bearing these points in mind, the agreement with observed values must be considered satisfactory.

Barth (1967) published similar data for the ultraviolet NI multiplets at 1200 Å $(^4S-{}^4P)$ and 1493 Å $(^2D-{}^2P)$. With 4 keV electrons he found the intensities to be 3.2 and 0.9 respectively, when normalized to the intensities in Table 5.1. Koval *et al.* (1969) compared auroral intensities in the 7000—1500 Å region with laboratory experiments on 150 eV and 13 keV electron impact on N_2 and air. The spectra are similar, but the 150 eV spectra agree somewhat better with the auroral spectrum than do the 13 keV spectra.

Ajello (1970) and Prasad and Green (1970) made similar studies in the ultraviolet, and found that direct dissociation of N_2 with excitation

Table 5.1 *Excitation rates by electron impact (from data of Stolarski and Green 1967)*

Excited state	Transitions	Wavelength in Å or band system[1]	Normalized excitation rates for primary electrons of various energies (keV). 30	10	1	0.1	Observed intensities (from Table 4.4)
$N_2^+ B^2\Sigma_u$	B→X	1 Neg.	165	165	165	165	165
$N_2^+ A^2\Pi_u$	A→X	Meinel	680	680	680	690	1500—4000
$N_2\ A^3\Sigma_u^+$	A→X	V-K	2610	2590	2570	4950	200
$N_2\ B^3\Pi_g$	B→A	1. Pos.	1270	1260	1250	2560	1000—5000
$N_2\ C^3\Pi_u$	C→B	2. Pos.	217	215	212	523	100— 800
$N_2 a^1\ \Pi_g$	a→X	L-B-H[2]	1300	1300	1380	2670	~100
$N_2\ b^1\ \Pi_u$	b→a b→X	Janin B-H	1060	1060	1120	2010	
$N_2\ h^1\ \Sigma_u^+$	h→a h→X	G-H W-K	590	600	630	1050	
$O_2^+ A^2\Pi_u$	A→X	2. Neg.	191	191	183	137	
$O_2^+ b^4\Sigma_g^-$	b→a	1. Neg.	157	156	146	110	80
$O_2\ a^1\Delta_g$	a→X	IR-Atmosph.	4650	4610	4940	7900	
$O_2\ b^1\Sigma_g^+$	b→X	Atmosph.	13	14	13	27	
$O\,^1D$	$\to {}^3P$	6300/64	930	920	880	1570	40— 200
$O\,^1S$	$\to {}^1D$	5577	96	94	97	200	200
	$\to {}^3P$	2972	6	6	6	12	
$O\,3s\,^5S^0$	$\to 2p^4\,^3P$	1356	92	92	93	240	10
$O\,3s\,^3S^0$	$\to 2p^4\,^3P$	1304	184	182	176	360	200
$O\,3p\,^5P$	$\to 3s\,^5S^0$	7773	4	4	4	11	10— 25
$O\,3p\,^3P$	$\to 3s\,^3S^0$	8446	86	76	71	143	20— 50
$O\,3d\,^3D^0$	$\to 2p^4\,^3P$	1027	10	10	11	27	
$O\,3s\,^3D^0$	$\to 2p^4\,^3P$	989	165	172	152	303	
$O\,3s\,^1D^0$	$\to 2p^4\,^1D$	1152	99	106	98	280	
$O\,3s\,^1P^0$	$\to 2p^4\,^1S$	1218	13	12	13	47	
$O\,3s\,^1P^0$	$\to 2p^4\,^1D$	998	25	24	26	50	

[1] V-K: Vegard-Kaplan. L-B-H: Lyman-Birge-Hopfield. B-H: Birge-Hopfield. G-H: Gaydon-Herman. W-K: Watson-Koontz.

[2] Cascading neglected.

is very important. The intensity distribution of the λ 1200 multiplet is strongly affected by resonance scattering (Strickland 1970).

Stolarski (1968) computed the detailed height emission profile for various lines and bands for a sample aurora. Though rather tentative, the results show that the height distribution of the first negative band, first and second positive bands, and the green [OI] line (λ 5577) should be rather similar, in agreement with observations.

Performing similar computations on the expected intensity ratio between the green [OI] line and the first negative bands, Omholt

(1959) also arrived at satisfactory agreement with observations, $I(5577)/I(4709)$ being about 25. Kamiyama (1966) likewise computed the $I(5577)/I(3914)$ ratio and found a value of 1.5. Dalgarno and Khare (1967), on the other hand, found this same ratio to be 0.4, assuming that collision with ambient, thermal electrons removed much of the available electron energy.

In all these computations various assumptions have been made on the "injection" of secondary electrons and on collision processes competing with the excitation of O^1S and O^1D for the available energy. Also, it should be noted that excitation of O^1S and O^1D has been assumed to be from atomic oxygen O^3P only. As pointed out, Parkinson *et al.* (1970) have recently cast serious doubt on the validity of this assumption, and have concluded that dissociation of O_2 is more important. This requires energies of about 9.3 and 7.1 eV respectively.

The most relevant cross-sections in the 0—8 eV energy range are shown in Fig. 5.1. These curves are based on data by Smith *et al.* (1967), Stolarski *et al.* (1967) and Watson *et al.* (1967). Since this figure is based partly on small-scale figures in the original papers, potential users of the data are encouraged to go back to these for accurate data. References to data on relevant cross-sections are given by Dalgarno *et al.* (1969) and in an appendix to this chapter. Semi-empirical cross-sections and relations between ionization yield, cross-sections and loss functions have been derived more recently by Peterson *et al.* (1969) and by Peterson and Green (1968).

Dalgarno *et al.* (1963) made computations on heating of the ambient electron gas by collisions with secondary electrons. They adopted a formula of Butler and Buckingham (1962). Above 3 eV the energy loss by electron-electron collisions may be expressed by

$$\left\{\frac{dE}{dx}\right\} = -1.95\cdot 10^{-12}\frac{\{n_e\}}{\{E\}} \tag{5.5}$$

where E is expressed in eV, n_e in cm^{-3} and x in cm. From detailed computations it appears that Eq. (5.5) holds with good accuracy also at 2 eV. From Fig. 8.2 it is seen that electron densities in aurora rarely exceed 10^6 cm^{-3}. At 4 eV this density gives an energy loss $-dE/dx$ of the order of 5×10^{-7} eV cm^{-1}. This may be compared to the energy loss by excitation of $O\,^1D$. With an atomic density of 10^{11} cm^{-3} (at about 120 km), the energy loss by excitation of $O\,^1D$ for 4 eV electrons is about 6×10^{-6} eV/cm^{-1}. This would indicate that energy loss by electron-electron collisions is not a dominating process in aurora, since the comparison is made with a fairly high electron density. However, the process may well be significant, in particular at greater altitudes. Dalgarno

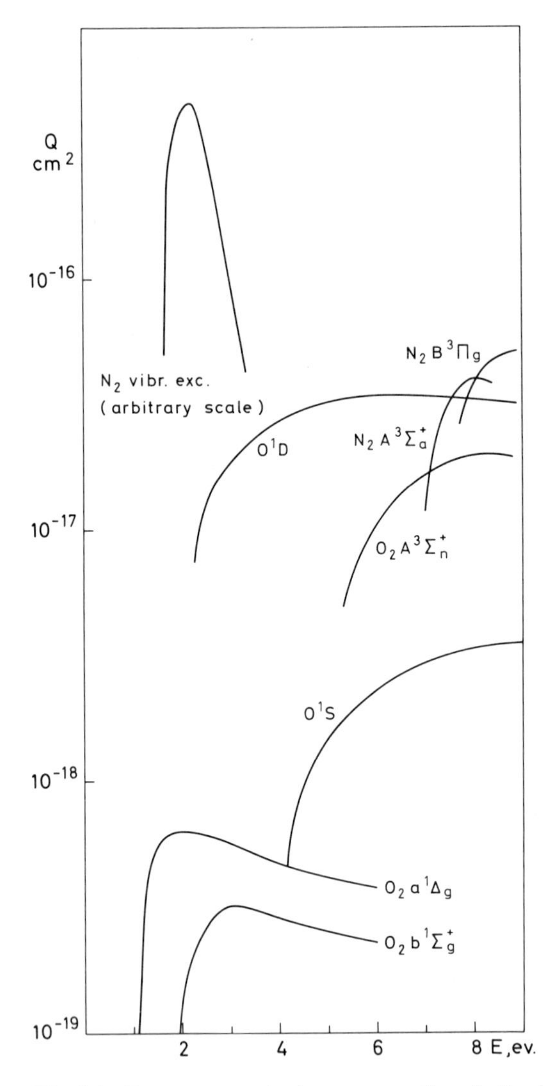

Fig. 5.1 Important excitation cross sections in the 1—8 eV range

and Khare (1967) and Rees *et al.* (1969) concluded that, because of this process, excitation by secondary electrons is not adequate to explain the excitation of low-lying levels, but there are a number of uncertainties in their basic data.

Vibrational excitation of N_2 has been studied in the laboratory by Haas (1957), Schulz (1959, 1962, 1964), Schulz and Koons (1966), Boness and Hasted (1966) and by Andrick and Ehrhardt (1966). Theoretical studies were made by Herzenberg and Mandl (1962) and Chen (1964). The excitation occurs through an intermediate formation of N_2^-, and the cross-section shown in Fig. 5.1 is that adopted by Green and Barth (1965). It is reasonably consistent with that adopted by Dalgarno *et al.* (1963) and with Chen's (1964) theoretical values. Vibrational excitation of N_2 will effectively remove all available energy in the 2—3 eV range, and hence reduce the excitation of O^1D. Vibrational excitation of O_2 and rotational excitation of molecules are probably uniportant (cf. Dalgarno *et al.* 1969, Dalgarno *et al.* 1963).

In conclusion, it may be stated that most theoretical evidence points towards excitation of atomic oxygen by 4—7 eV electrons as a main source for excitation of $O\,^1D$ and $O\,^1S$. There are, however, certain uncertainties involved, and there is evidence (Parkinson *et al.* 1970) that simultaneous dissociation of O_2 and excitation is also an important process. Similarly, dissociation of N_2 may be an important source for NI-lines.

In any case, the rate of excitation of $O\,^1S$ and $O\,^1D$ relative to the rate of ionization seems to be only a little influenced by the height of the aurora and the incident energy of the primary particles (Stolarski 1968, Parkinson *et al.* 1970). Since the excitation of the first negative N_2 bands is strictly related to the rate of ionization, the intensity ratio $I(5577)/I(3914)$ is also expected to be constant.

5.1.2 Thermal Collisions

Dissociative recombination

$$X\,Y^+ + e \rightarrow X + Y \tag{5.6}$$

is believed to be an important reaction in the upper atmosphere. When dissociative recombination of O_2^+, N_2^+ and NO^+ occurs, energy is available for excitation of the lowest lying states. Table 5.2 shows available energy and possible excited states.

Table 5.2 *Data relevant to dissociative recombination*

Molecule	Ionization energy	Dissociation energy[1)]	Excess energy	Possible excited atoms
	eV	eV	eV	
N_2	15.5	9.7	5.8	$N(^2D)$, $N(^2P)$
O_2	12.5	5.1	7.4	$O(^1D)$, $O(^1S)$
NO	9.5	6.5	3.0	$O(^1D)$,[2)] $N(^2D)$

[1)] Gaydon 1950.
[2)] $N(^4S)+O(^1D)$ as dissociation products violate spin conservation and are therefore less likely (Dalgarno and Walker 1964).

Approximate recombination coefficients are given in Table 5.3 (from data collected by Biondi 1969).

Table 5.3 *Approximate coefficients for dissociative recombination (in units of* 10^{-7} cm^3 s^{-1}*)*

Molecule	T=200 °K	400 °K	800 °K
N_2^+	3	2.5	2
O_2^+	3	1.7	1
NO^+	7	3	1.5

Considerable uncertainties have attached to the efficiencies for such processes to produce the excited atoms in question. Recent experimental data by Zipf (cf. Mukherjee 1970) give factors of 0.10 and 0.90 for production of $O(^1S)$ and $O(^1D)$ from O_2^+ by dissociative recombination. Dissociative recombination of NO^+ is probably unimportant for production of $O(^1D)$ because of spin conservation rules. A comprehensive discussion of these and other excitation processes is given by Mukherjee (1970).

Dalgarno and Khare (1967) considered dissociative recombination an important source for $O(^1S)$ in aurora. However, from direct rocket measurements of λ 5577 emission and ion densities, Parkinson *et al.* (1970) concluded that this process contributes significantly at high altitudes only, and perhaps supplies about 10 per cent of the total λ 5577 emission. This agrees with the fact that time-delaying excitation mechanisms cannot be traced in the measurements of rapid intensity variations of λ 5577 and the first negative N_2^+ bands (cf. Sect. 5.4). Stoffregen (1970) has invoked dissociative recombination to explain an apparent enhancement of the λ 5577 emission from an aurora below an artificial barium cloud. In view of what is said above, this appears less likely.

Dissociative recombination as a source for the red oxygen lines 6300/64 emitted by $O(^1D)$ has been considered by Rees *et al.* (1967) and by Mukherjee (1970). With a 0.9 efficiency factor for the process, it seems able to contribute about 10—30%, depending on height.

Direct, radiative or three-body recombination of atomic ions is probably of little importance. Recombination of atomic ions in the upper atmosphere takes place mostly through charge transfer to molecules, followed by dissociative recombination.

Charge transfer processes may contribute to excitation of various atoms and molecules. However, little is known about their efficiency and few, if any, are definitely proved to be important. The general ion chemistry in the ionosphere during aurora is still far from being completely understood and rocket measurements have revealed surprising data (cf. Donahue *et al.* 1970, Swider and Narcisi 1970).

Malville (1959) has considered the process

$$N_2^+ + O^-(\text{or } O_2^-) \rightarrow N_2(B\,^3\Pi) + O(\text{or } O_2) \tag{5.7}$$

as a possible source for the first positive N_2 bands, particularly in type B aurora (red lower border). This process appears also to have been proposed by Mitra (1943) and by Gosh (1943).

Gadsden (1961, 1962) has proposed a similar mechanism which may contribute to the excitation of the second positive N_2 bands and the red [OI] lines:

$$N_2^+ + O^- \rightarrow N_2(C^3\Pi) + O(^1D). \tag{5.8}$$

Two processes with fairly close energy resonance are (Omholt 1957):

$$O^+(^2P) + O_2(X\,^3\Sigma_g^-) \rightarrow O(^3P) + O_2^+(b\,^4\Sigma_g^-), \tag{5.9}$$

and

$$O^+(^2D \quad \text{or} \quad ^2P) + N_2(X^1\,\Sigma_g^+) \rightarrow O(^3P) + N_2^+(A^2\,\Pi). \tag{5.10}$$

In the latter process there is almost exact energy resonance with $O^+(^2D)$ and $N_2^+(A\,^2\Pi, v'=1)$.

Various processes involving deactivation of metastable states have been proposed as important for transferring energy from one kind of particle to another. For such processes to be important, enough energy must be available; energy resonance and spin conservation are also considered to be important.

Seaton (1954) pointed out that there is a close energy resonance in the process

$$O(^1D) + O_2(X\,^3\Sigma_g^+) \rightarrow O(^3P) + O_2(b\,^1\Sigma_g^+, v=2). \tag{5.11}$$

Since vibrational energy of O_2 is removed in collisions with other particles, the reverse process is not important. Wallace and Chamberlain (1959) conclude from studies of the intensity distribution among O_2 atmospheric bands that process (5.11) is the most important one for excitation of these bands.

Meyer *et al.* an (1969) found a rate coefficient of $3 \times 10^{-12}\,\text{cm}^3\,\text{s}^{-1}$ for the process

$$N_2(A\,^3\Sigma_u^+) + O(^3P) \rightarrow N_2(X\,^1\Sigma_g^+) + O(^1S). \tag{5.12}$$

This would make the contribution from this process significant but not dominating at high altitudes.

The process

$$N(^2D) + O(^3P) \rightarrow N(^4S) + O(^1D) \tag{5.13}$$

is important, being one of the most important processes for deactivating $N(^2D)$ atoms. These have a lifetime of about 26 hours, so that very little energy is radiated directly (cf. Sect. 5.2). On the basis of this, as much as half of the λ 6300 excitation may be due to this process at altitudes between 200 and 300 km.

A similar process, deactivating $O^+(^2D)$, was proposed by Makadevan and Roach (1968):

$$O^+(^2D) + O(^3P) \rightarrow O^+(^4S) + O(^1D) . \tag{5.14}$$

This process, too, was found to be important by Mukherjee (1970) but at greater altitudes only. Above 400 km it may contribute as much as 30 per cent, while it is insignificant below 200 km (cf. Gerard 1970).

Those processes which involve ions or long-lived metastables as one of the collision partners lead to time-delay effects in the excitation. Measurements of rapid time variations in the auroral spectrum should give a clue as to the extent to which thermal collisions involving ions or metastables do excite other auroral emissions. The point is that in all cases an appreciable time (about 0.5 s and upwards) will elapse, on the average, from the time when the ion or metastable particle is produced until the energy transfer takes place. If the production of these ions and metastables is in phase with the primary ionization, a time lag should occur between the latter and the emissions excited by thermal collisions. With photoelectric techniques such a time lag should be observable in rapidly varying aurora.

Eftestøl and Omholt (1965) discussed and measured this effect for excitation of the Meinel and first negative N_2^+ bands. Assuming as an hypothesis that the first negative bands are excited by primary and secondary electrons only, they found that thermal collisions probably contribute to excitation of the other emissions with a minor, but significant fraction. The results must be considered as tentative only.

Evans and Vallance Jones (1965) made a similar study of the time lag between the first negative N_2^+ bands and the first positive N_2 bands and found that the possible time lag was less than 0.1 sec. Studies on the green and red oxygen lines are consistent with the view that energy transfer collisions are important in exciting λ 6300/64, but not for λ 5577 (cf. Sect. 5.4).

It was early suspected by Vegard (cf. Vegard *et al.* 1955) that the sodium doublet at 5890/96 Å $(3s\,^2S - 3p\,^2P)$ sometimes is enhanced in aurora, as compared to the nightglow level. Hunten (1955) similarly suspected that it was enhanced in Type B aurora. It was conclusively proved that this was the case by Derblom (1964). Direct excitation of sodium atoms by primary particles and secondary electrons seems inefficient, because of the low density of sodium atoms. Hunten (1965) suggested that sodium is excited through

$$\mathrm{Na}+\mathrm{N}_2(v\geq 8)\rightarrow \mathrm{Na}(^2\mathrm{P})+\mathrm{N}_2. \qquad (5.15)$$

The cross-section for excitation of N_2 to the $v=8$ level by electrons with energies between 2.6 and 3.0 eV is of the order of 10^{-17} cm^2 (Chen 1964). Comparing this with the cross-sections displayed in Fig. 8.1 and with Stolarski's (1968) computations of various excitation rates, a serious guess on the excitation rate of $N_2(v=8)$ would be that it is of the same order of magnitude as that of the green oxygen line by direct excitation of atomic oxygen, or perhaps somewhat higher. The intensity of the sodium doublet is about one per cent of that of the green line (1 kR in an aurora of brightness coefficient III). Since the fraction of sodium atoms in air is only about 10^{-10}, sodium atoms must be 10^7 to 10^8 times as effective as other particles in de-populating the $v=8$ level of N_2. Hunten (1965) considered a reaction rate coefficient of 3×10^{-10} cm^3 s^{-1} or even larger to be possible, while Zipf (1969) considered 10^{-10} cm^3 s^{-1} as an appropriate order of magnitude. However, the rate coefficient for quenching by other N_2 molecules is about 10^{-14} to 10^{-15} cm^3 s^{-1} for $v=4-10$ (Zipf 1969), and combination of these values does not permit sufficient excitation of sodium. However, all data are uncertain, and the process considered here should not be disregarded. In any case it serves excellently for illustrative purposes.

5.1.3 Excitation by Thermal Electrons

It was pointed out in Sect. 5.1.1 that the secondary electrons also lose energy by collisions with the electrons in the ambient electron gas. The rate of energy loss is given by Eq. (5.5). This raises the temperature of the electron gas above that of the ions and neutrals, such that the ambient electrons may contribute to the excitation of optical emissions. To compute the electron temperature one has to consider the balance between the heating process and the loss processes, the latter being mainly collisions with the ion and neutral gases and thermal conduction.

The heating of the electron gas by photoelectrons in the daytime and the resulting diurnal variation in electron temperature have been extensively studied (cf. Dalgarno *et al.* 1967). In aurora the problem is more complicated because of the complicated geometry. Rees *et al.* (1967) made computations on an aurora observed over Alaska, with brightness coefficient about III. Between 120 and 400 km they found the electron temperature to be about 3 times that of the neutral gas, and the ion gas was also significantly heated. At high altitudes (800 km) the electron and ion temperature were almost equal and about 4500 °K, compared to 1200 °K for their neutral model atmosphere. The heating rate of the electrons was of the order 0.1 to 1 eV s^{-1} per electron, so that the equilibrium time for a temperature of a few thousand degrees must be of the order of one second. Similar results, but with somewhat lower temperatures, were obtained by Stolarski (1968). There is also clear evidence from rocket data (cf. McNamara 1969, Baker *et al.* 1967, Holt and Lerfald 1967) that the electron temperature greatly exceeds that of the main atmosphere.

With their model, Rees *et al.* (1967) found that excitation of the red [OI] lines by thermal electrons is insignificant at low altitudes, increasing to about 10% of the total rate of excitation at 200 km and becoming dominant above 300 km (where the emission is still relatively significant because of deactivation at lower altitudes). Similar results were obtained by Mukherjee (1970), but with a contribution of about 30 to 40 per cent between 500 and 800 km. This lower fraction was due to the large contribution from process (5.14), discussed in the preceding section.

Heated, thermal electrons may also be important in excitation of the infrared atmospheric bands of O_2 (Megill *et al.* 1970). The excitation potential of the a $^1\Delta_g$ state is only 1.0 eV.

5.1.4 Discharge Mechanisms and Heating by Electric Fields

Electric discharges have from time to time been advocated as an important source of excitation in aurora (cf. Chamberlain 1961). This theory has been rejected mainly on the following basis: in a discharge the intensity distribution varies greatly with atmospheric density for a given current or for a given electric field strength. With vertical discharges in the aurora the current must be approximately constant with height, and with horizontal discharges the electric field strength must be constant (because the conductivity is high along the magnetic field lines). There are no drastic height variations in the main auroral spectrum which can be attributed to such mechanisms. Considerations of the likely electric field strengths in aurora and comparison with earlier computations by Chamberlain (1955) led to the conclusion that discharges were of little importance in the aurora as well as in the night glow, while it was shown that an electric field may enhance the ion temperature considerably (Omholt 1959, 1962). Mukherjee (1970) also concluded that discharges were unimportant in excitation of the red oxygen lines (exc. pot. 2 eV) up to 800 km.

Rees and Walker (1967) made detailed computations on the heating of electrons and ions by electric fields. Their computations on two model auroras with maximum electron densities of 2.6×10^6 cm^{-3} and 1.1×10^5 cm^{-3} at 150 and 220 km respectively give the following results: with an electric field of 0.05 V m^{-1} the electron temperature rises 100—300 °K above what it would otherwise be, while the ion temperature rises by 1000—2000 °K. For a field of 0.01 V m^{-1} heating is unimportant for electrons as well as for protons. Since the latter seems more typical for the field within the luminous aurora, we can probably conclude that heating by electric fields usually contributes very little to the excitation. This also holds for the larger of the two field strengths, but in that case the contribution to the red oxygen line should not be completely neglected at great heights. An exception may be the $O_2(a^1\Delta_g)$ state, which has an excitation potential of 1.0 eV only. Megill *et al.* (1970) and Noxon (1970a) found that the intensity of the infrared atmospheric bands required an efficiency of at least 10 per cent in converting primary particle energy into excitation of $O_2(a^1\Delta_g)$ and in extreme cases the required efficiency even exceeded 100 per cent. It is not inconceivable that an electric field of 0.05 to 0.1 V m^{-1} suffices to provide the additional energy required.

5.2 Deactivation

Deactivation of excited atoms and molecules by collisions with other atmospheric species is of considerable importance in the upper atmosphere. Energy is transferred from one species of particles to another, causing the intensities of emission lines to be changed.

If the probability of collisional deactivation is unity per gas-kinetic collision, then the total probability of deactivation is equal to the collision frequency. In the upper atmosphere this ranges from about 10^5 s^{-1} at 80 km to about 10^2 s^{-1} at 120 km and about 1 s^{-1} at 300 km. This must be compared to the spontaneous transition probabilities, which are 10^5 to 10^8 s^{-1} for excited, ordinary states and about 10^{-4} to 1 s^{-1} for the metastable states in question. Hence, for allowed transitions, quenching by collisional deactivation is unimportant, whereas only a small probability of deactivation per gas-kinetic collision may be significant for metastable species. With a probability of one per gas-kinetic collision, the reaction rate coefficient (the probability per quenching particle in one cm^3) is about 2—3 $\times 10^{-10}$ cm^3 s^{-1} at 300 °K.

Table 5.4 gives a review of the metastable atoms and molecules of particular importance in aurora together with emissions, radiative lifetimes, probable collision partners in important quenching collisions, and deactivation rate coefficients. It is seen that in a few cases the rate coefficient approaches the gas kinetic value. The values quoted are weighted averages of experimental data, and mainly collected from an important review paper by Zipf (1969). The last entry gives the approximate height h_1 in the atmosphere where deactivation by collision is as probable as spontaneous radiation of a photon. Hence, at some scale-heights below this height not much radiation will be found, since quenching has become dominant. The computations were made on the basis of CIRA 1965 Atmospheric Model 5 (average solar activity) at midnight. For values above 150 km, h_1 may vary considerably with solar activity and also to a certain extent in the course of a day. For example, these variations together make h_1 for the 6300 [OI] line vary between the extreme values of 250 and 400 km.

Many of the values for deactivation rate coefficients should be regarded as tentative only. The differences between values given in an earlier review paper by Hunten and McElroy (1966) and those quoted here and by Zipf (1969) are often large. Also present data are partly divergent, which is apparent from the experimental data listed by Zipf (1969). Laboratory experiments to measure these rate coefficients are often ambiguous, and therefore, as Hunten (1969) pointed out, rocket probe measurements of atmospheric emissions may provide a valuable supplement to laboratory data. In the upper atmosphere nature provides experiments under ideal conditions—low pressure and no walls—hence unwanted or unknown secondary effects are minimized.

Table 5.4 *Lifetimes and deactivation probabilities for metastable species in aurora*

Metastable species	Emissions	Lifetime (radiative)	Ref.	Important collision partner	Probable deactivation rate coeff.	Ref.	h_1 [1)]
	λ in Å	seconds			($cm^3\ sec^{-1}$)		km
$O(^1S)$	5577 and 2972	0.74	1,2	O_2 O	$1-3\times10^{-13}$ 2×10^{-13}	6,7	95
$O(^1D)$	6300 and 6364	110	1	N_2 O_2	8×10^{-11} (10^{-12}) $3-6\times10^{-11}$	6,8 14 6.8	300 (170)
$O(^5S)$	1356	6×10^{-4}	1				
$O^+(^2P)$	7319, 7330 and 2470	5	1				
$O^+(^2D)$	3726 and 3729	1.3×10^4 (3.6 h)	1	N_2 O	3×10^{-10} $10^{-12}-10^{-10}$	6,9 15	500
$N(^2P)$	10395, 10404 and 3466	12	1	N_2			
$N(^2D)$	5198 and 5201	9.10^4 (26 h)	1	O_2	$1-2\times10^{-12}$	6,10	300
$N^+(^1S)$	5755 and 3063	0.9	1				

Table 5.4 *Lifetimes and deactivation probabilities for metastable spezies in aurora (cont.)*

Metastable species	Emissions	Lifetime (radiative)	Ref.	Important collision partner	Probable deactivation rate coeff.	Ref.	h_1 [1)]
	λ in Å	seconds			($cm^3 sec^{-1}$)		km
$N^+(^1D)$	6548 and 6584	250	1				
$N_2(A^3\Sigma_u^+)$	Vegard-Kaplan bands	1.3 (F_2) 2.6 (F_1, F_3)	3	O N	3×10^{-11} 5×10^{-11}	6,11 6	140
$O_2(a^1\Delta_g)$	Infrared atmospheric bands	2700	4	N O_2^-	3×10^{-13} 2×10^{-10}	6,12	
$O_2(b^1\Sigma_g^+)$	Atmospheric bands	12	5	N_2	1.5×10^{-15}	6,13	90

[1)] h_1 is the height at which collisional deactivation equals spontaneous transitions (CIRA 1965; Model 5; 00 h).

References to Table 5.4

1. Garstang 1951, 1952, 1956 and 1961.
2. Experimental from aurora cf. Sect. 5.4.
3. Shemansky and Carleton 1969, Shemansky 1969, Broadfoot and Maran 1969. Earlier arguments were partly in favour of longer lifetimes, about 10—15 sec, cf. Brömer and Spieweck 1967.
4. Badger, Wright and Whitlock 1965.
5. Childs and Mecke 1931, Wark and Mercer 1965, Burch and Gryvnak 1967, Cho, Allin and Welsh 1963.
6. Zipf 1969.
7. Zipf 1967, Young and Black 1966, Stuhl and Welge 1969, Black *et. al.* 1969.
8. Snelling and Bair 1967, Hunten and McElroy 1966, Young, Black and Slanger, cf. Zipf 1969, Noxon, cf. Zipf 1969, Parkinson *et. al.*, cf. Zipf 1969, De More and Raper 1964a, 1966, Wallace and McElroy 1966, Peterson and Vanzandt 1969.
9. Dalgarno and McElroy 1966, Stebbings, Turner and Rutherford 1966, Banks, cf. Zipf 1969.
10. Hunten and McElroy 1966, Hernandez and Turtle 1969.
11. Hunten and McElroy 1966, Young, cf. Zipf 1969.
12. Hunten and McElroy 1968.
13. Young and Black 1967, Wallace and Hunten 1968, Noxon, cf. Zipf 1969, Izod and Wayne, cf. Zipf 1969, Ogryslo 1969, Stuhl and Welge 1969.
14. Derived from measurements by Murcray (1969). Murcray (1969) has measured the intensity ratio $I(6300)/I(5577)$ as function of height by rocket photometers. From his data a deactivation coefficient of about 10^{-12} cm^3 s^{-1} is derived (his coefficient $\alpha=10^{-10}$ cm^3 is equal to d_2/A_3, probability, and not to d_2, as erroneously assumed in Murcray's comparison with airglow data).
15. Suggested by Mahadevan and Roach (1968).

5.3 Interpretation of the Spectrum

There is no question of the dominating importance of fast particles and secondary electrons in the excitation of the auroral spectrum. For most lines and band systems there is satisfactory agreement between observed and theoretical intensities, considering the present uncertainties in the data on basic processes and in the observed intensities. Table 5.1 shows reasonable agreement between theory and observations except for those emissions which are strongly forbidden and therefore heavily quenched. One important exception is the Lyman-Birge-Hopfield band system of N_2. Another is the infrared atmospheric O_2 band system, which may be excited through heated, thermal electrons or by discharge mechanisms (cf. Sects. 5.1.3 and 5.1.4).

In this section we shall go into some detail on the interpretation of the intensity distribution, and in particular we shall discuss atomic lines and the intensity distribution within the molecular band systems.

5.3.1 Atomic Lines

The forbidden [OI] lines $\lambda\,5577\,(^1D-{}^1S)$ and $\lambda\,6300/64\,(^3P-{}^1D)$ are among the most frequently discussed and observed auroral emissions. From the discussions in Sects. 5.1 and 5.2 (also anticipating the discussion in Sect. 5.4) it is reasonably certain that excitation of λ 5577 is performed mainly through direct excitation of O or dissociation and excitation of O_2, by primary particles and secondary electrons. Dissociative recombination may contribute at greater heights, probably with less than 10 per cent of the total emission, integrated over height. Also transfer of energy from $N_2(A\,^3\Sigma_u^+)$ by process (5.12) may occur. Deactivation is of minor importance except at unusually low heights.

With the λ 6300/64 multiplet the situation is different. Collisional deactivation is very efficient at heights where most other emissions are strongest (cf. Sect. 5.2). Hence the λ 6300/64 emission is mainly located above 200—300 km (cf. Sect. 4.2.3). Although secondary electrons are important, thermal energy transfer collisions and excitation by heated, thermal electrons together probably give a much larger contribution (cf. Sect. 5.1).

The intensity ratios between some other atomic oxygen lines are not so readily explained, (cf. also Hicks and Chubb 1970). It is seen from Table 4.4 that the intensity ratio between the $^3P-{}^5S$ transition at 1356 Å and the $^3P-{}^3S$ transition at 1304 Å is about 0.05, whereas Stolarski and Green's (1967) computations yield a value of about 0.5, a factor of ten higher. On the other hand, the observed intensity ratio between the $^5S-{}^5P$ transition at 7773 Å and the $^3S-{}^3P$ transition at 8446 Å is about 0.5 as compared to a theoretical value of about 0.05, a factor of ten lower.

The former of the two intensity ratios may be influenced by strong imprisonment of the λ 1304 radiation, since the $2p^4\,{}^3P-3s\,{}^3S$ is a resonance transition, and by deactivation of the metastable $3s\,{}^5S$ state. Since the intensity data are from satellite observations, performed from above, the observed intensity of the λ 1304 radiation may be up to twice as much as it would be without the imprisoning effect (Hicks and Chubb 1970). This is because the emission radiated downwards is absorbed and to a large extent re-emitted upwards. The measured λ 1356/λ 1304 ratio is also not very accurate, and in one case it was found to be about 0.2 (Barth and Schaffner 1970). Since neither of the theoretical values is very accurate, there is not necessarily any inconsistency present as regards this ratio. The fact that the $3s^5S$ state is metastable probably does not affect the intensity ratio. The lifetime is less than 10^{-3} s (Garstang 1961), and an excessively high quenching efficiency would be necessary to affect the intensity. It should be noted, however, that the present intensities (cf. Table 4.4) indicate that the dominating excitation process for the λ 1356 multiplet is cascading from $3p\,^5P$ through the λ 7773 multiplet. Comparing the observed and computed intensities (cf. Table 4.4), it appears that the theoretical computations do not predict the intensity of the λ 1304 multiplet too badly, whereas the λ 7773 and λ 8446 multiplets are out by factors of 2—5 in opposite directions, and the λ 1356 multiplet is too small by a factor of ten. It remains to be investigated whether direct dissociation and excitation of O_2 may explain the observations.

Particular interest attaches to lines from O^+ and N^+, and whether these, in addition to the N_2^+ bands, may serve as a measure of ionization and dissociation rates during an aurora. The transitions observed with certainty are listed in Table 4.2. Excitation is probably performed through direct ionization of atoms as well as dissociation and ionization of molecules.

Chamberlain (1961) has estimated the sum of the intensities of permitted NII and OII lines in aurora of brightness coefficient III to be about 50 and 10 kR respectively, and those for the forbidden lines to be about 1 and 5 kR respectively. This must be compared to an intensity of the order 100 kR in the first negative N_2^+ bands and a total rate of ionization in a vertical column of the order of 2×10^{12} ion pairs $cm^{-2}\,s^{-1}$. Because of cascading, not all the photons are due to an excitation process which at the same time is an ionization process. Hence, the emission from N^+ and O^+ ions represents only a few percent of the total rate of ionization.

Comparison between the cross-section for total ionization of N_2 by electrons and that for production of N_2^+ indicates that about one fifth of the ionization of N_2 leads to production of ionized atoms rather than to molecular ions (cf. Dalgarno *et al.* 1965). It is likely that most of these atomic ions are produced in the ground configurations $2p^2$ and $2p^3$ for nitrogen and oxygen respectively. This requires energies of the order of 30 eV or less, whereas excitation of higher configurations requires an additional 20 eV or more.

A sizeable fraction of the ions are probably left in the excited, metastable states of the $2p^2$ and $2p^3$ configurations, but collisional deactivation is likely to induce considerable quenching. The lifetimes of the $O^+(^2P)$ and $N^+(^1S)$ states are about 5 and 1 sec, respectively, and of the $O^+(^2D)$ and $N^+(^1D)$ states about 4 hours and 4 minutes, respectively (cf. Table 5.3). Gerard (1970) has made an attempt to compute the O^+ ion densities and the intensities and height variations of the λ 7319/30 ($^2D-{}^2P$) and λ 3729 ($^4S-{}^2D$) [O II] multiplets. The agreement with observations was satisfactory for the $^2D-{}^2P$ multiplet, but not for the $^4S-{}^2D$ multiplet, which came out with too little intensity in the computation. However, as pointed out by Gerard (1970) much depends on the flux of soft primary particles, since the emission mainly occurs at great heights.

Particular observations and discussions of the [NII] line at 6584 Å ($^3P_2-{}^1D$) have been published by Harang and Pettersen (1967) and Belon and Clark (1959), of the NII line at 5003 Å ($3s\,^3P_0-3p\,^3S_1$) by Cherednichenko (1964), Vaysberg (1962) and Ivanchuk (1961), and of the NII line at 4176 Å ($3d\,^1D-4f\,^1F$) by Gerard and Harang (1969) (cf. also Omholt 1970). The λ 5003 line was particularly related to proton impact, being correlated with the Hβ intensity (Ivanchuk 1961). The excitation of some other NII multiplets have been studied in the laboratory by Dufay, Desequelles, Druetta and Eidelsberg (1966).

During aurora the concentration of atomic nitrogen and oxygen will increase due to dissociative recombination, which follows ionization of N_2 and O_2. Brown (1968) has considered the enhancement of the atomic nitrogen concentrations during aurora, and finds that for typical densities of auroral ionization the nitrogen concentration rapidly rises to levels in excess of daytime conditions. However, he finds that the rate of excitation due to electron impact on atomic nitrogen even with the enhanced atomic nitrogen concentration, is less than that due to dissociative recombination of molecular ions. Maeda and Aikin (1967) made a study of dissociation of O_2 due to the impact of auroral particles, and find that it may perhaps be of some importance during quiet auroras with a hard electron energy spectrum. We may conclude that dissociation does not usually alter the atmospheric composition to any degree of importance for the auroral spectrum.

5.3.2 The Intensity Distribution within Molecular Band Systems

The intensity distribution among the various vibrational bands in a molecular emission band system is given (cf. e.g. Nicholls 1962) by:

$$I_{v'v''}(\mathrm{BA}) = C_{\mathrm{BA}} N_{v'}(\mathrm{B}) E^3_{v'v''} p_{v'v''} \tag{5.16a}$$

$$= C_{\mathrm{BA}} N_{v'}(\mathrm{B}) E^3_{v'v''} R_e^2(\bar{r}_{v'v''}) q_{v'v''} \tag{5.16b}$$

Here the intensity I is measured in emitted photons (not energy) per unit number of molecules and time, v' and v'' denote the vibrational quantum numbers in the upper (B) and lower (A) state respectively, $E_{v'v''}$ is the photon energy, $R_e^2(\bar{r}_{v'v''})$ is the electronic transition moment, and $q_{v'v''}$ the Franck-Condon factor. $R_e^2(\bar{r}_{v'v''})$ is a slowly varying function of $\bar{r}_{v'v''}$, the average internuclear distance during the transition:

$$\bar{r}_{v'v''} = \int \psi_{v'} r \psi_{v''} \mathrm{d}r \,/ \int \psi_{v'} \psi_{v} \mathrm{d}r \,. \tag{5.17}$$

The Franck-Condon factor $q_{v'v''}$ is given by

$$q_{v'v''} = \left| \int \psi_{v'} \psi_{v''} \mathrm{d}r \right|^2 , \tag{5.18}$$

where ψ_v is the vibrational wave function.

$$p_{v'v''} = R_e^2(\bar{r}_{v'v''}) q_{v'v''} \tag{5.19}$$

is the so-called band strength.

Except for metastable states with long lifetimes, the energy distribution is generally undisturbed by collisions. The rate of population of a particular vibrational level is then

$$g(v') = \sum_{v''} I_{v'v''} = C_{\mathrm{BA}} N_{v'}(\mathrm{B}) \sum_{v''} p_{v'v''} E_{v'v''}^3 \,. \tag{5.20}$$

If the $p_{v'v''}$'s are known, the rate of population may be derived from one single band, since substituting $\mathrm{N}_{v'}(\mathrm{B})$ from Eq. (5.16a) gives

$$g(v') = I_{v'v''} \cdot \frac{\sum\limits_{v''} p_{v'v''} E_{v'v''}^3}{p_{v'v''} E_{v'v''}^3} \,. \tag{5.21}$$

If the excitation is performed by electrons, the number of excitations from the ground state X and vibrational level v''' to the B, v' level is given by (cf. e.g. Nicholls 1962):

$$I_{v'''v'}(\mathrm{XB}) = C_{\mathrm{XB}} N_{v'''}(\mathrm{X}) G_{\mathrm{XB}}^2(\bar{r}_{v'''v'}) q_{v'''v'}(\mathrm{XB}) \,. \tag{5.22}$$

Here G_{XB}^2 is the transition moment, again slowly dependent on the average internuclear distance. For optically allowed transitions G_{XB}^2 is usually assumed to be of the same form as R_e^2.

Since the transition moments R_e^2 and G^2 vary slowly with $\bar{r}$, and hence with vibrational quantum number, the intensity distribution among the various bands and the excitation rates to the various vibrational levels depends largely on the Franck-Condon factor q. Nicholls (1962) and Zare, Larson and Berg (1960) have given Franck-Condon factors for a number of band systems. Recent measurements of the Vegard-Kaplan bands have been made by Chandraiah and Shepherd (1967).

With reasonable temperatures (cf. Chapter 6) almost all molecules in the ground states are in the zeroth vibrational level. Hence the rates of excitation are not strongly dependent on the temperature of the atmosphere.

With the relevant basic data at hand it is possible to compute theoretical, synthetic spectra and compare with observations. Also the finite slit width of the spectrograph must be taken into account in the computations. This is a powerful way of analyzing the more complex parts of the auroral spectrum. An example of such an analysis, made by Shemansky and Vallance Jones (1968) is shown in Fig. 5.2.

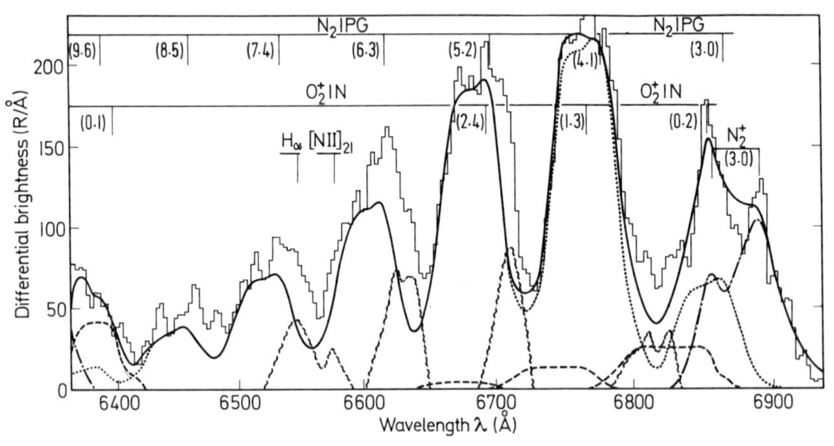

Fig. 5.2 Spectrum of type B aurora.
Total synthetic spectrum ——— N_2 1 PG .... (250 °K), N_2^+ Meinel —·—·—, O_2^+ 1 N ———, Difference Spectrum ∼∼∼. $\Delta\lambda = 15$ Å. (From Shemansky and Vallance Jones, 1968, courtesy Pergamon Press)

When computing the actual rates of population one must also take into account cascading from higher states. For example, the $B^3\Pi_g$ state of N_2 is populated both by direct excitation from the ground state and through cascading from the $C^3\Pi_u$ state by emission of the second positive bands (cf. Table 4.3).

The pioneer work in analysis of auroral band systems was done by Bates (1949a, b), who analyzed the first negative N_2^+ bands and the first and second positive N_2 bands. No account was then taken of variations of R^2 and G^2 with $\bar{r}$. Since then much work has been done, and we shall cite the most important papers which have appeared during the decade 1960—70. Earlier work, up to 1960, has been cited by Chamberlain (1961).

Broadfoot and Hunten (1964) studied the Vegard-Kaplan bands and the first and second positive bands of N_2. They concluded that nearly all the excitation of the Vegard-Kaplan bands occurred through cascading (first positive bands). Ahmed (1969) arrived at the same conclusion. This is not consistent with Stolarski and Green's (1967) calculations (cf. Table 5.1). Also, they concluded that cascading through the second positive bands contributed appreciably to the excitation of the first positive bands, by about 17%. Shemansky and Vallance Jones (1968) found significant differences in the vibrational distribution in the first positive bands between type B aurora and high-level aurora. About 5% contribution from cascading fitted the data best, but in no case very accurately (cf. Sect. 4.2.5). They also made some tentative studies of the O_2^+ first negative bands, finding no disagreement between predicted (from electron excitation) and observed intensities. Also the first negative N_2^+ bands have an intensity distribution consistent with electron excitation (cf. Chamberlain 1961).

Benesch *et al.* (1966, 1967) concluded that the vibrational development of the second positive N_2 bands indicated that the vibrational temperature of the molecules in the ground state is about 2000°K, in which case higher vibrational levels in the ground state become populated. The vibrational temperature could possibly be raised by electron excitation (cf. Sect. 5.1.1). For the first positive bands their conclusion on cascading is similar to that of Shemansky and Vallance Jones (1968).

There was some experimental evidence from the laboratory that excitation by protons leads to a higher vibrational development than excitation by electrons (cf. Chamberlain 1961). Such evidence is not unambiguous, since with dense excitation sources pure vibrational excitation may also be operative, and later investigations fail to detect any enhancement (Philpot and Hughes 1964, Sheridan and Clark 1965). There is no evidence from auroral spectra that any particular vibrational enhancement occurs in proton aurora (cf. Gerard and Harang 1970).

In conclusion, we may say that the intensity distributions observed are consistent with the assumption that the dominating excitation process is electron and proton excitation from the ground state of neutral molecules, in reasonable thermal equilibrium. At high altitudes, there is perhaps a different vibrational distribution due to vibrational excitation of molecules in the ground state, and little collisional redistribution to thermal equilibrium. One must bear in mind that the auroral emissions originate from a great range of heights with several orders of magnitude difference in density. The existing disagreements between observation and theory are probably not greater than may be accounted for by observational errors and imperfections in the theoretical work. Thus, for example, in theoretical work variations in the excitation moment $G^2(\bar{r}_{v'v''})$ have not been seriously taken into account.

Important exceptions to the particle excitation may be the excitation of the atmospheric O_2 bands (cf. Sect. 5.1.2, Eq. 5.11), and the infrared atmospheric O_2 bands (cf. Sect. 5.1.4).

5.4 The Lifetime of Metastable Oxygen Atoms

The green oxygen line at 5577 Å is emitted from the ^1S level in oxygen, which has a lifetime against radiation of about 0.7 s. This means that in aurora of rapidly varying intensity the emission of this line, from one particular volume of air, should occur later than that of permitted radiations, such as the first negative N_2^+ bands, because emission is delayed compared to excitation. Once one is aware of it, this effect can easily be seen in auroras. Narrow rays which move rapidly across the sky show a violet leading edge and a green dusk behind. This effect has been used to study the lifetime of the O^1S atoms in the upper atmosphere (Omholt and Harang 1955, Omholt 1956, 1959, Paulson and Shepherd 1965, Evans and Vallance Jones 1965). That such measurements are possible was also proposed independently by Swings (1948).

Records demonstrating the effect are shown in Figs. 5.3 and 5.4.

The population n of the ^1S level in oxygen is given by

$$\frac{\mathrm{d}n}{\mathrm{d}t} = Q-(A_{32}+A_{31}+d_3)n=Q - \frac{1}{\tau}n\,, \tag{5.23}$$

where Q is the rate of excitation, A_{32}, A_{31} and d_3 are respectively the probabilities of spontaneous transitions to the ^1D level (by emission of λ 5577), to the ^3P levels (by emission of λ 2972) and of collisional deactivation of ^1S atoms. τ is the effective lifetime of oxygen atoms in the ^1S level, given by

$$\frac{1}{\tau} = A_{32}+A_{31}+\mathrm{d}_3\,. \tag{5.24}$$

The emission of λ 5577 photons is given by

$$I_{\mathrm{O}} = A_{32}n\,. \tag{5.25}$$

The emission of photons in the first negative bands happens within $10^{-7}-10^{-8}$ sec, hence for our purpose we may always consider the emission of these bands to be proportional to, and in phase with, the excitation. The fact that the intensity ratio between the first negative bands and the green line in quiet aurora usually is reasonably constant (cf. Sects. 4.2.2 and 5.3.1) lends support to the view that the corresponding

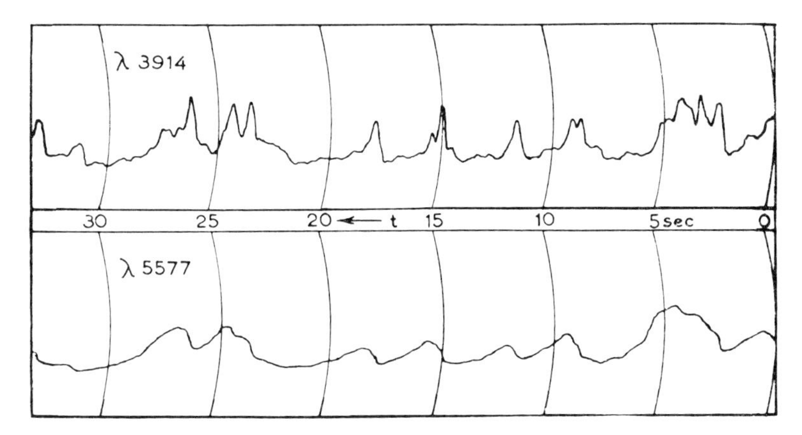

Fig. 5.3 Record of flashing aurora, Yerkes Observatory.

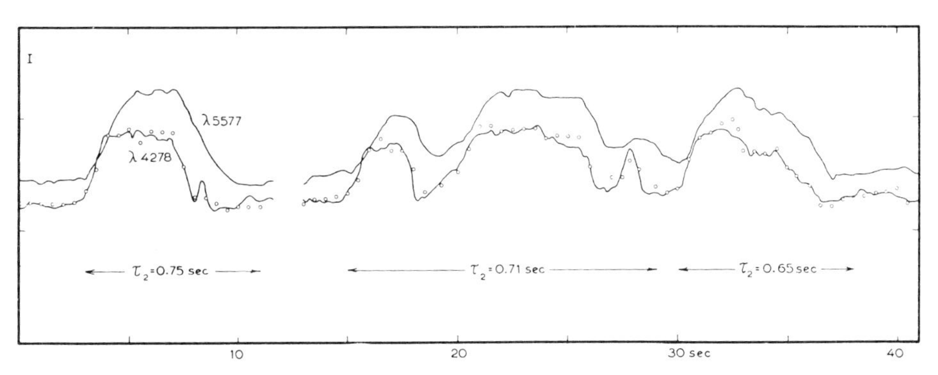

Fig. 5.4 Record of pulsating aurora, Tromsø. The intervals indicated by ⟷ were treated as individual records from which the given values of τ were computed. ooo: Curve for $I(4278)$ constructed from the curve for $I(5577)$ by Eq. (5.27), taking $\tau = 0.75$ s

excitation rates are approximately proportional. Adopting this, i.e.

$$Q = k\, I_{\mathrm{N}}, \tag{5.26}$$

we get from Eqs. (5.23) and (5.25)

$$K\, I_{\mathrm{N}} = I_{\mathrm{O}} + \tau \frac{\mathrm{d} I_0}{\mathrm{d} t}, \tag{5.27}$$

where

$$K = k\, A_{32}\, \tau\,. \tag{5.28}$$

Hence, detailed measurements of I_N, I_O and dI_O/dt in rapidly varying auroras should furnish data for computing K and τ. It is essential for the accuracy that the intensity variations are rapid enough to make $\tau \, dI_O/dt$ of the same order of magnitude as I_O and KI_N.

There are three main methods which have been used for finding the constants K and τ from observations. The first is a graphic method which was employed by Omholt and Harang (1955) and by Omholt (1956) and later also by Evans and Vallance Jones (1965). From the records of I_N and I_O, these quantities as well as dI_0/dt can be derived simultaneously at different points along the curve, or as average values for different intervals to increase accuracy. Eq. (5.27) now describes a straight line in a K, τ coordinate system. Hence, each set of measurements describes a straight line, and provided Eq. (5.27) describes the observed phenomenon accurately, these should all cross at one point, giving the true values of K and τ. Actually, there is a scatter of crossing points between individual lines, indicating that Eq. (5.27) is not rigorously valid. In most cases, however, it seems to hold reasonably well, and fairly accurate values of K and τ can be derived.

A second method is that of least squares, adopting as the best values of K and τ those which, inserted in Eq. (5.27) with the measured intensities, yield a minimum value for

$$\Delta^2 = \int_{t_1}^{t_2} \left(K I_N - I_O - \tau \frac{dI_O}{dt} \right)^2 dt \tag{5.29}$$

(Omholt 1959). If the intensities and their derivatives are measured at discrete intervals, the corresponding finite sum of deviations should be minimized. The magnitude of $\Delta^2_{\min}$ is a measure of the validity of Eq. (5.27).

An important disturbing factor in measurements of weak optical signals is noise, which is partly (and mostly) shot noise due to the statistical nature of a current being generated by photons, and partly instrument noise. Since the noise is almost white, i.e. is distributed evenly at all frequencies, whereas the auroral variations are mostly confined to a limited frequency region, methods involving frequency analysis are superior to others. Such methods may consider, or put most weight on, the signal in only those frequency regions where the signal-to-noise ratio is highest.

The above-mentioned methods are simple and require little computing, but with the advent of electronic computers this is no longer a serious problem. A method based on frequency analysis was developed by Paulson and Shepherd (1965). Since their method is rather sophisticated, and there is no evidence that this is an advantage, we shall here illustrate the point by a simpler and more direct approach.

Consider a linear relation between two time-dependent quantities $X(t)$ and $Y(t)$ and their derivatives:

$$Y(t)+\beta\frac{\mathrm{d}}{\mathrm{d}t}Y(t)=aX(t)+b\frac{\mathrm{d}}{\mathrm{d}t}X(t). \tag{5.30}$$

If the Fourier spectra of these quantities are described by

$$X(t)=\int_0^\infty x(\omega)\mathrm{e}^{\mathrm{j}(\omega t+\phi_x)}\mathrm{d}\omega \tag{5.31}$$

and

$$Y(t)=\int_0^\infty y(\omega)\mathrm{e}^{j(\omega t+\phi_y)}\mathrm{d}\omega, \tag{5.32}$$

where the Fourier amplitudes $x(\omega)$ and $y(\omega)$ are real, then it is easy to show, by inserting in Eq. (5.30), that

$$y(\omega)=\left(\frac{a^2+\omega^2 b^2}{1+\omega^2\beta^2}\right)^{\frac{1}{2}}x(\omega), \tag{5.33}$$

and the phase difference

$$\phi(\omega)=\phi_y-\phi_x=\tan^{-1}(\omega\beta)-\tan^{-1}\left(\frac{\omega b}{a}\right). \tag{5.34}$$

The coefficients a, b and β may be determined from the experimental curves for $y(\omega)/x(\omega)$ and $\phi(\omega)$. If $\beta=0$, as is assumed in this case, the relations are even simpler. We then obtain from Eq. (5.33):

$$i_\mathrm{N}(\omega)=\frac{1}{K}(1+\omega^2\tau^2)^{\frac{1}{2}}i_\mathrm{O}(\omega) \tag{5.35}$$

and from Eq. (5.34):

$$\tan\phi(\omega)=-\omega\tau, \tag{5.36}$$

where $i_\mathrm{N}(\omega)$ and $i_\mathrm{O}(\omega)$ are the Fourier amplitudes of I_N and I_O.

Also, the following relation holds:

$$\frac{i_\mathrm{N}(\omega)\cos\phi(\omega)}{i_\mathrm{O}(\omega)}=\frac{1}{K}. \tag{5.37}$$

By Fourier analysis of the two intensities, and by plotting the ratio $i_\mathrm{N}/i_\mathrm{O}$ or the phase angle ϕ as function of ω, the lifetime τ may be determined; in the first case also the intensity ratio K. If the basic equation (5.27) is correct, then Eq. (5.37) should also hold for all values of ω. Since the observation and instrument noise shows a rather flat spectrum and random phases, whereas auroral observations are peaked at a lower frequency, one will eventually come to frequencies where Eq. (5.37) is not valid. This may provide information as to the frequencies at which auroral signals exceed the noise level.

Paulson and Shepherd (1965) analysed simultaneous records of the N_2^+ λ 3914 band and the [OI] λ 5577 line on this basis, but with a somewhat different mathematical approach. They computed the auto-correlation functions ρ_N and ρ_O and the cross-correlation functions ρ_{NO} of the two intensities, as well as the Fourier spectra for these. The phase shift between the Fourier components of ρ_{NO} and ρ_O is as given by Eq. (5.36), and the "coherency" function

$$\gamma(\omega) = \frac{f_{NO}(\omega)}{(f_N(\omega)\, f_O(\omega))^{\frac{1}{2}}}, \tag{5.38}$$

$f(\omega)$ being the Fourier amplitudes for the correlation functions, should be unity provided the basic relation between I_N and I_O holds. The auroral signal dominates over noise up to the frequency where $\gamma(\omega)$ begins to deviate appreciably from unity. Also, the relation between f_{NO} and f_N is the same as between i_O and i_N, except for normalizing factors.

All observations give strong support to the view that K is constant to a good approximation during the few seconds of a record used for computing K and τ. Paulson and Shepherd (1965) found the coherence function γ to be very close to one, up to a frequency limit which ranged from 0.3 to 1 s^{-1} ($\nu = \omega/2\pi$), and their plot of tan ϕ versus ω gave straight lines up to almost the same limit. In a few cases the data for the higher values of ω indicated a higher value of τ than those for the lower ones.

Since the intensities often were measured on arbitrary scales, values of K are not absolute nor comparable. The derived values of τ are shown in Table 5.5 With the likely errors in the individual results taken into account, the data are in good agreement with a radiative lifetime of about 0.70—0.75 s for the $O\,^1S$ atoms, and that deactivation occurs in the lower auroras. Since values as low as 0.45—0.50 s undoubtedly are real, the deactivation coefficient d_3 in Eq. (5.24) must sometimes be as large as 0.7 s^{-1}. This must be an average value over the height interval covered by the actual aurora, and is in good agreement with the data given in Table 5.4. According to these a deactivation coefficient of about 1.4 s^{-1} is to be expected at about 95 km, whereas most auroras occur somewhat above this height. For pulsating auroras, deactivation coefficients of about 0.2—0.3 should not be uncommon. The measured lifetime and constancy of the factor K are consistent with the view that excitation processes which would give large time-delays in the excitation, such as dissociative recombination of O_2^+ and energy transfer from

Table 5.5

Auroral form	Author	τ (seconds)
Draperies and bands	(Omholt and Harang 1955)	0.45 – 0.8
	(Omholt 1956)	0.4 – 0.85
	(Omholt 1959)	0.5 – 0.8
	Average of 15 records:	0.68 ± 0.11
Type B	(Evans and Vallance Jones 1965)	0.48 ± 0.06
Pulsating aurora	(Omholt and Harang 1955)	0.55 – 0.6
	(Omholt 1956)	0.6 – 0.75
	(Omholt 1959, two events	0.74 ± 0.10
	with several records each)	0.57 ± 0.06
Pulsations in "quiet" auroral forms	(Paulson and Shepherd 1965)	0.47 – 0.73 + 0.03
Flaming aurora	(Evans and Vallance Jones 1965)	0.67 ± 0.06

$N_2(A\,^3\Sigma_u^+)$ through process (5.12), are of minor importance (cf. Parkinson *et al.* 1970, Parkinson and Zipf 1970).

Attempts have also been made to make measurements of the effective lifetime of the $O(^1D)$ atoms in aurora from irregular time variations in the auroral intensity, using graphical and least square methods (Stoffregen and Derblom 1960, Omholt 1960). In this case apparent lifetimes ranging from -10 to $+525$ s were found, the average value being 190 s and the median 140 s. Subsequent photometer records (Omholt, unpublished) of the 6300 line and N_2^+ bands revealed that Eq. (5.27) in most cases fitted the data rather poorly, and the conclusion was that the lifetime of $O\,^1D$ atoms cannot be measured this way.

This is not surprising, considering the known facts. The deactivation coefficient for $O(^1D)$ atoms by O_2 and N_2 is two orders of magnitude higher than that for $O(^1S)$ atoms by O and O_2 (cf. Table 5.4). On the other hand, the lifetime against radiation is about 110 s. This results in quenching of λ 6300 radiation below about 200 km. Hence the observed N_2^+ and λ 6300 emissions from aurora originate at vastly different heights, and the observed λ 6300 emission is excited mostly (directly or indirectly) by the extreme low energy particles among the primaries and by secondary energy transfer processes (cf. Sect. 5.1.2). This, together with the fact that a record used to measure the lifetime of the 6300 line must last 5—10 min, leaves little possibility of the ratio between the excitation rates remaining constant throughout the record. Also, because observations from the ground integrate over height and the lifetime varies with height from almost zero to perhaps its full, undisturbed value, a measured average value would be of small physical significance.

Actually, the long apparent lifetimes measured may indicate that the red line excitation in the highest part of the aurora is delayed compared to the main excitation in the lower part. This is in agreement with the view that energy transfer processes are important (cf. Sect. 5.1.2).

Despite this difficulty more regular, simultaneous pulsations in the λ 6300 and λ 5577 intensities have been measured (Eather 1969). Periods of even less than 10 sec were observed, although the modulation of the λ 6300 intensity was very small, less than one per cent. This was found to be consistent with the general knowledge about the height distribution of auroral luminosity and the quenching coefficient. The modulated λ 6300 light comes from the lower part of the auroral emission region, with an appreciable decrease in effective lifetime of the $O\,^1D$ atoms. From the time lag between the two intensity curves and relative damping of the 6300 oscillations, an estimate could be made of the quenching rate in the lower part of the aurora. Eather (1969) found this rate to increase as the pulsation period decreased, and found that the only reasonable explanation was an overall lowering of the effective height of the λ 6300 emission as the period decreases. This agrees with Johansen's (1966) conclusion from a comparison of pulsation frequency and radio wave absorption efficiency of pulsating auroras (cf. Sect. 6.9).

5.5 Helium Emissions

If α-particles are among the primary particles incident on the atmosphere during aurora, one would expect helium lines to be emitted, due to excitation mechanisms similar to those which are responsible for the hydrogen lines. Fan (1956), on the basis of laboratory experiments, suggested that the multiplet at 5876 Å ($1s2p\,^3P-1s3d\,^3D$) might be observable in aurora, provided about 10% or more of the primaries were α-particles.

Eather (1966) made detailed computations on the expected emission of the λ 5876 multiplet, with α-particles of various energies as primaries. If an originally pure He^{++} beam is incident on the atmosphere, equilibrium will quickly be established between He^{++}, He^{+}, and He, due to charge exchange processes, similarly as for protons (cf. Sect. 3.2). Laboratory data show that the fraction of He^{++} left in the beam reaches equilibrium already at a height of about 300 km. Hence, in the detailed computations, which are very similar to those for a proton-hydrogen beam, only He^{+} and He need be considered. One should note that the ground state in helium is a ^1S state, whereas the easiest observable emissions, the λ 5876 ($1s2p\,^3P-1s3d\,^3D$) and λ 10830 ($1s2s\,^3S-1s2p\,^3P$) multiplets, belong to the triplet system. Transition between the two sets of states, parahelium (singlets) and orthohelium (triplets), occurs only by electron exchange or through ionization followed by recombination.

Electron exchange can be neglected in the case of fast helium atoms colliding with atmospheric species.

The equilibrium fractions of ionized, para- and orthohelium can in principle be computed, together with the rate of excitation of the various excited states in question. Due to uncertainties in the cross-sections, there are large uncertainties in the calculated emission intensities. For 250 keV α-particles impinging upon a nitrogen atmosphere, the lower and upper limits for the λ 5876 emissions were found to be respectively 10 and 31 photons per α-particle. With equal fluxes of protons and α-particles at energies above 200 keV, the intensity of λ 5876 should be about twice that of $H\beta$ (Eather 1966).

That emissions from helium atoms occasionally occur in aurora has been suspected by various authors. Bernard (1947, 1948) tentatively identified a number of features in the auroral spectrum as due to helium, but the identifications were not confirmed. Mironov *et al.* (1958, 1959) found a spectral feature which coincided with the λ 10830 multiplet. This emission was detected in the famous great red aurora of 11 February, 1958. Also Shefov (1961, 1963) and Fedorova (1961, 1962) found evidence for He lines in the auroral spectrum.

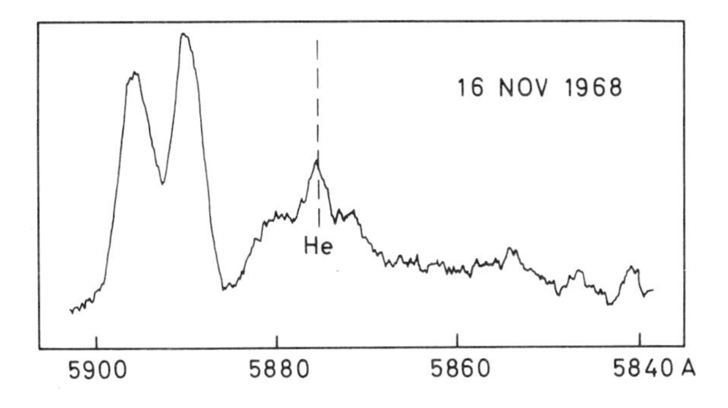

Fig. 5.5 The helium λ 5876 multiplet in aurora (from records by Stoffregen 1969)

Conclusive evidence that helium lines may occur seems to have been brought forward by Stoffregen (1969), who observed the λ 5876 multiplet. Fig. 5.5 shows some of his observations, when blending by rotational OH lines or other features is not prohibitively strong. Such blending has also been the main obstacle in attempts to observe the λ 10830 multiplet.

It appears from Stoffregen's observations that the intensity of the λ 5876 multiplet reached values from 50 to 120 R, and that the multiplet was observed in short periods of strongly enhanced intensity only.

A typical case is shown in Fig. 5.6. That the helium lines are observable only for short periods seems to be consistent with earlier observations and the fact that most observations fail to show helium lines (Eather 1967).

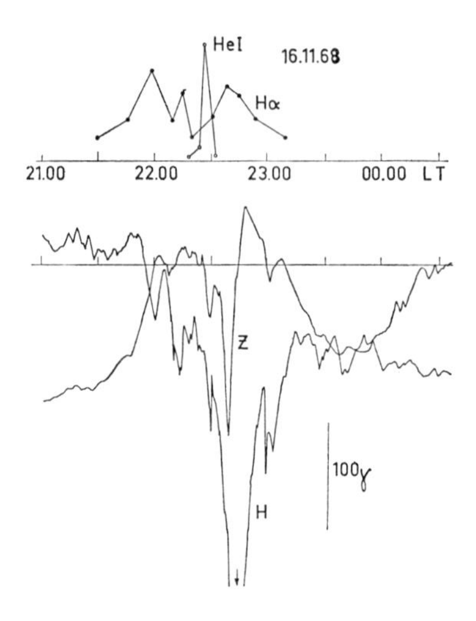

Fig. 5.6 The appearance of the helium λ 5876 multiplet and Hα on the 16. Nov. 1968. Below: geomagnetic activity. (From records by Stoffregen 1969, courtesy Pergamon Press)

Reasoner *et al.* (1968) made a coordinated experiment with rocket-borne particle detectors and ground-based photometers. In that particular aurora (9 Feb. 1967) the ratio of differential flux of primary α-particles with energy $\sim$200 keV to protons with energy $\sim$100 keV was about 1.6×10^{-2}. Simultaneous measurements from the ground on the $H\alpha$ line and the λ 5876 He multiplet gave an intensity of about 50 R for $H\alpha$ and an upper limit of 1 R for the helium emission. The $H\alpha$ intensity was found to be consistent with the proton fluxes measured at various energies, and the absence of helium emission indicated that the flux of α-particles was less than 2% of the proton flux. This, they maintain, is in agreement with the most common value for the α-particle-to-proton ratio in the solar wind. This is about 2—3%, although values up to 15% occasionally occur.

Stoffregen's (1969) measurements, compared to Eather's (1966) computations, would indicate that the flux of α-particles on rare occasions may be comparable to the proton flux. However, with the uncertainties involved, the data are not necessarily inconsistent with an upper limit of about 15 per cent, which is inferred from satellite measurements (Hundhausen *et al.* 1967).

Harrison and Cairns (1969) conclusively observed a relatively strong emission of the helium multiplet λ 10830 in a sunlit aurora. The intensity was about 15 kR in an aurora of brightness coefficient II, in a fully illuminated atmosphere. Their observation confirmed earlier work by Fedorova (1962), who reported that this helium multiplet may be observed in sunlit aurora but not in the dark (cf. also Shefov 1963).

The λ 10830 emission is a resonance multiplet from the lowest state of the triplet series, $1s2\,^3S$. This is a strongly metastable state, and for thermal helium atoms transition to the ground state ($1s^2\,^1S$) occurs only through collisional deactivation. A detailed treatment of the production and destruction of orthohelium in the upper atmosphere is given by McElroy (1965). With an atmospheric maximum temperature of 1500 °K the lifetime of He ($1s2s\,^3S$) atoms is about 7 sec at 300 km and 80 sec at 500 km, and with a temperature of about 750 °K the lifetimes are 20 sec and 400 sec respectively. Metastables are produced through recombination from ions and electron exchange collisions with photoelectrons (produced by solar radiation). They are destroyed through ionization and deactivating collisions with ambient atoms and molecules. The excitation potential of the metastables is nearly 20 eV, so that ionization of the collision partner may occur, which makes the collisional cross-sections relatively high.

The most plausible explanation for the occurence of the λ 10830 multiplet in sunlit auroras is excitation of the ambient He ($1s^2\,^1S$) atoms in the atmosphere to the metastable $ls2s\,^3S$ state by secondary electrons, the latter being produced by the primaries (Shefov 1961, 1963, Harrison and Cairns 1969). The metastables then resonantly scatter the λ 10830 solar radiation, in a similar way to the N_2^+ ions (cf. Sect. 5.6). Hence, the enhanced helium emission in sunlit auroras is due to the thermal helium atoms in the uppermost part of the atmosphere, contrary to the sporadic helium emissions observed in ordinary auroras, which are due to neutralized, incident α-particles. Rough computations by Harrison and Cairns (1969) on the expected ratio between the λ 10830 multiplet and the green oxygen line λ 5577 give the right order of magnitude.

5.6 Sunlit Aurora

Størmer (1955) found that aurora in the sunlit atmosphere (sunlit aurora) is situated much higher than ordinary aurora in the dark atmosphere. The lower border of sunlit auroral rays is mostly at 200—250 km, compared to about 150 km for rays in the dark. Auroral arcs in the sunlit atmosphere show a similar effect. Another most remarkable phenomenon is the divided auroral rays observed by Størmer. On some occasions an auroral ray disappears on the shadow-line but re-appears again further down, in the dark atmosphere, and along the same magnetic field line. Spectra of sunlit auroras show strong enhancement of the first negative N_2^+ bands compared to the green oxygen line. The visual colour of sunlit auroras is blue and grayish, compared to yellow-green for ordinary auroras. More recently, the helium emission at 10830 Å has also been identified in sunlit aurora (cf. Sect. 5.5).

An explanation of the enhancement of the N_2^+ bands was offered by Bates (1949a, b) in terms of resonance scattering of sunlight from an enhanced density of N_2^+ ions, produced by the primary particles. According to this theory, a sunlit aurora is composed of a weak, possibly subvisual, high-altitude ordinary aurora with a resonance scattering aurora superimposed. A divided auroral ray then consists of an upper part where resonance scattering makes the aurora visible, an intermediate, non-sunlit part which is subvisual, and a lower part which is intense enough to be visible.

In sunlit auroras both the vibrational and rotational distribution is enhanced compared to ordinary auroras. Vallance Jones and Hunten (1960) found rotational temperatures of the order of 2200 °K and vibrational temperatures of the order of 2000 °K. These values may be regarded as typical. Ahmed (1969) found the rotational temperatures to vary from about 1800 to 2600 °K. These temperatures may not be far from the gas kinetic temperatures (cf. Chamberlain 1961).

Rees (1959) and Lytle and Hunten (1960) derived densities of N_2^+ in sunlit aurora from the intensities of the N_2^+ bands and theoretical calculations of the resonance scattering. They found maximum ion densities of the order of 10^4 to 10^5 cm^{-3}.

Broadfoot (1967) made a thorough study of the resonance scattering by N_2^+. He also took into account the production of N_2^+ ions in excited states from the ground state of N_2. This, as well as the resonance scattering process itself, strongly affects the vibrational distribution of the N_2^+ ions in the ground state. Detailed calculations on the vibrational development as function of true temperature and ion lifetime, and of the total scattering effect led Broadfoot to the surprising conclusion that only 40 per cent of the first negative band emission is resonance scattered. As he pointed out, this conclusion is inconsistent with the earlier understanding of sunlit aurora, and it also yields a lifetime of the ions of about 0.25 sec, which is unexpectedly short. A lifetime of some tens of seconds against recombination would be more appropriate at these heights ($\sim$400 km).

Obviously more elaborate computations and careful observations, of the Meinel N_2^+ bands, too, are needed to resolve all the details.

However, it seems likely that it will ultimately be possible to explain most of the emission by resonance scattering together with particle ionization and excitation.

Appendix

References to Data on Cross-Sections.

A general discussion of important cross-sections is given by Dalgarno *et al.* (1969).

O: Ionization: Fite and Brackman (1959), Rothe *et al.* (1962). Excitation: ^1D and ^1S: Smith *et al.* (1967), $3s\,^3S$: Stauffer and McDowell (1966).

O_2: Ionization: Tate and Smith (1932), Graggs *et al.* (1957), Lampe *et al.* (1957), Fite and Brackmann (1959), Rapp and Englander-Golden (1965) and Schram *et al.* (1965, 1966). Simultaneous ionization into the $b\,^4\Sigma_g^-$ state of O_2^+: Stewart and Gabathuler (1958), Nishimura (1966, 1968), and Aarts *et al.* (1968). Dissociative attachment: Rapp and Briglia (1965). Inelastic scattering of electrons with energies between 4.5 and 12.5 eV: Schultz and Dowell (1962) McGowan *et al.* ((1964), Hake and Phelps (1967). Inelastic scattering of electrons at energies in the region of 500 eV: Lassettre *et al.* (1964), Silverman and Lassettre (1964).

N_2: Ionization: Tate and Smith (1932), Lampe *et al.* (1957), Rapp and Englander-Golden (1965), and Schram *et al.* (1965, 1966). Ionization with excitation of $B\,^2\Sigma_u^+$: Stewart (1956), Sheridan *et al.* (1961), Hayakawa and Nishimura (1964), Davidson and O'Neil (1965), McConkey and Latimer (1965), McConkey *et al.* (1967), Holland (1967), Hartman (1968), Srivastava and Mirza (1968), Nishimura (1968), and Aarts *et al.* (1968). Ionisation and excitation of $A\,^2\Pi_g$: Zapesochnyi and Skubenich (1966), Srivastave and Mirza (1968). Excitation: $B\,^3\Pi_g$: Williams (1935), Zapesochnyi and Skubenich (1966), Simpson and McConkey (1969), Shemansky and Broadfoot (1970), Engelhardt *et al.* (1964) have obtained an estimate of the cross-section for a transition occurring near 6.7 eV, which probably corresponds to excitations of the $B\,^3\Pi_g$ and $A\,^3\Sigma_u^+$ states (Takayanagi and Takahashi 1966). $C\,^3\Pi_u$: Thieme (1932), Langstroth (1934), Herrmann (1936), Stewart and Gabathuler (1958), Kishko and Kuchinka (1959), Zapesochnyi and Kishko (1959), Fink and Welge (1964), Zapesochnyi and Skubenich (1966), Jobe *et al.* (1967), Burns *et al.* (1969). $a\,^1\Pi_g$: Holland (1968), Ajello (1970). Takayanagi and Takahashi (1966) have derived from the high-velocity data the Born approximation to the cross-sections for excitation of the $a\,^1\Pi_g$ state, the $b\,^1\Pi_g$ state, and a group of states of N_2 near 14 eV. Dissociation: Winters (1966). Dissociation and excitation:

Ajello (1970). Inelastic scattering of electrons with energies in the region of 500 eV: Lassettre and Krasnow (1964), Silverman and Lassettre (1964), Geiger and Stickel (1965), Lassettre *et al.* (1965), and Meyer and Lassettre (1966).

References

Aarts, J. F. M., de Heer, F. J., Vroom, D. A.: Physica **40,** 197 (1968).
Ahmed, M.: J. Atmospheric Terrest. Phys. **31,** 1259 (1969).
Ajello, J. M.: J. Chem. Phys., **53,** 1156 (1970).
Andrick, D., Ehrhardt, J.: Z. Physik **192,** 99 (1966).
Badger, R. M., Wright, A. C., Whitlock, R. F.: J. Chem. Phys. **43,** 4345 (1965).
Baker, K. D., Pfister, W., Ulwick, J. C.: Space Research VII. North Holland Publ. Co. p. 665 (1967).
Barth, C. A.: The Birkeland Symposium on Magnetic storms and Aurora. (Ed. A. Egeland and J. Holtet, CNRS) 1967.
— Schaffner, S.: J. Geophys. Res. **75,** 4299 (1970).
Bates, D. R.: Proc. Roy. Soc. (London) Ser. **A 196,** 217 (1949a).
— Proc. Roy. Soc. (London) Ser. **A 196,** 562 (1949b)
Belon, A. E., Clark, K. C.: J. Atmospheric Terrest. Phys. **16,** 220 (1959).
Benesch, W., Vanderslice, J. T., Tilford, S. G.: J. Atmospheric Terrest. Phys. **28,** 431 (1966).
— — — J. Atmospheric Terrest. Phys. **29,** 251 (1967).
Bernard, R.: Compt. Rend. **225,** 352 (1947).
— Rept. Garriot Com. p. 93. The Physica Society, London 1948.
Biondi, M. A.: Can. J. Chem. **47,** 1711 (1969).
Black, G., Slanger, T. G., St. John, G. A., Young, R. A.: Can. J. Chem. **47,** 1872 (1969).
Boness, M. J. W., Hasted, J. B.: Phys. Lett. **21,** 526 (1966).
Broadfoot, A. L.: Planetary Space Sci. **15,** 1801 (1967).
— Hunten, D. M.: Can. J. Phys. **42,** 1212 (1964).
— Maran, S. P.: J. Chem. Phys. **51,** 678 (1969).
Brown, R. R.: J. Atmospheric Terrest. Phys. **30,** 55 (1968).
Brömer, H. H., Spieweck, F.: Planetary Space Sci. **15,** 689 (1967).
Burch, D. E., Gryvnak, D. A.: Philco-Ford Corp. Aeronutronic Div. U-4076 (1967).
Burns, D. J., Simpson, F. R., McConkey, J. W.: J. Phys. B (Proc. Phys. Soc.) **2,** 52 (1969).
Butler, S. T., Buckingham, M. J.: Phys. Rev. **126,** 1 (1962).
Chamberlain, J. W.: In The Airglow and the Aurorae. (Ed. E. B. Armstrong and A. Dalgarno) Pergamon Press 1955.
— Physics of the Aurora and Airglow. Academic Press 1961.
Chandriaiah, G., Shepherd, G. G.: Can. J. Phys. **46,** 221 (1967).
Chen, J. C. Y.: J. Chem. Phys. **40,** 3513 and **41,** 3263 (erratum) (1964).
Cherednichenko, V. I.: Geomagnetizm i Aeronomiya **4,** 456 (Engl. ed.) (1964).
Childs, W. H. J., Mecke, R.: Z. Physik **68,** 344 (1931).
Cho, C. W., Allin, E. R., Welsh, H. L.: Can. J. Phys. **41,** 1991 (1963).
CIRA: Cospar International Reference Atmosphere 1965.
Craggs, J. D., Thorburn, R., Tozer, B. A.: Proc. Roy. Soc. (London) Ser. **A 240,** 473 (1957).

Dalgarno, A.: In Atomic and Molecular Processes. (Ed. D. R. Bates) Academic Press 1962.
— Khare, S. P.: Planetary Space Sci. **15,** 938 (1967).
— Latimer, I. D., McConkey, J. W.: Planetary Space Sci. **13,** 1008 (1965).
— McElroy, M. B.: Planetary Space Sci. **14,** 1321 (1966).
— — Moffett, R. J.: Planetary Space Sci. **11,** 463 (1963).
— — Walker, J. C. G. Planetary Space Sci. **15,** 331 (1967).
— — Stewart, A. I.: J. Atmospheric Sci. **26,** 753 (1969).
— Walker, J. C. G.: J. Atmospheric Sci. **21,** 463 (1964).
Danilov, A. D.: Chemistry of the Ionosphere. (Engl. transl) Plenum Press 1970.
Davidson, G., O'Neil, R.: Proc. Fourth Intern. Conf. of Phys. of Electronic and Atomic Collisions 1965.
DeMore, W. B., Raper, O. F.: Astrophys. J. **139,** 1381 (1964a).
— — J. Chem. Phys. **44,** 1780 (1966).
Derblom, H.: J. Atmospheric Terrest. Phys. **26,** 791 (1964).
Donahue, T. M., Zipf, E. C., Parkinson, T. D.: Planetary Space Sci. **18,** 171 (1970).
Dufay, M., Desequelles, J., Druetta, M., Eidelsberg, M.: Ann. Geophys. **22,** 614 (1966).
Eather, R. H.: J. Geophys. Res. **71,** 4133 (1966).
— In The Birkeland Symposium on Aurora and Magnetic Storms. (Ed. A. Egeland and J. Holtet. CNRS) 1967.
— J. Geophys. Res. **74,** 4998 (1969).
Eftestøl, A., Omholt, A.: Geofys. Publikasjoner **25,** no 6 (1965).
Engelhardt, A. G., Phelps, A. V., Risk, G. G. Phys. Rev. **135,** A 1566 (1964).
Evans, W. F. J., Vallance Jones, A.: Can. J. Phys. **43,** 697 (1965).
Fan, C. Y.: In The Airglow and the Aurorae. (Ed. E. B. Armstrong and A. Dalgarno) Pergamon Press 1956.
Federova, N. I.: Planetary Space Sci. **5,** 75 (1961).
— Izv. Akad. Nauk SSSR, Ser. Geofiz. **4,** 538 (1962).
Fink, V. E., Welge, K. H.: Z. Naturforsch. **19 A,** 1193 (1964).
Fite, W. L., Brackmann: Phys. Rev. **113,** 815 (1959).
Gadsden, M.: J. Atmospheric Terrest. Phys. **22,** 105 (1961).
— J. Atmospheric Terrest. Phys. **24,** 750 (1962).
Garstang, R. H.: Monthly Notices Roy. Astron. Soc. **111,** 115 (1951).
— Astrophys. J. **115,** 506 (1952).
— In The Airglow and The Aurorae. (Ed. E. B. Armstrong and E. Dalgarno) London: The Pergamon Press Ltd., pp 324—327 (1956).
— Proc. Cambridge Phil. Soc. **57,** 115 (1961).
Gaydon, A. G.: Dissociation Energies. Dover Publ. Co. 1950.
Geiger, J., Stickel, W.: J. Chem. Phys. **43,** 4535 (1965).
Gerard, J. C.: Private comm. (1970).
— Harang, O.: Planetary Space Sci. **17,** 1680 (1969).
— — Phys. Norvegica **4,** 217 (1970).
Gilmore, F. R., Bauer, E., McGowan, J. W.: J. Quant. Spectr. Radiative Transfer. **9,** 157 (1969).
Gosh, S. N.: Proc. Natl. Inst. Sci. India **9,** 301 (1943).
Green, A. E. S., Barth, C. A.: J. Geophys. Res. **70,** 1083 (1965).
— Dutta, S. K.: J. Geophys. Res. **72,** 3933 (1967).
Haas, R.: Z. Physik **148,** 177 (1957).
Hake, R. D., Phelps, A. V.: Phys. Rev. **158,** 70 (1967).
Harang, O., Pettersen, H.: Planetary Space Sci. **15,** 1599 (1967).
Harrison, A. W., Cairns, C. D.: Planetary Space Sci. **17,** 1213 (1969).

Hartman, P. L.: Report LA-3793. Los Alamos Scientific Laboratory, Univ. of California 1968.
Hayakawa, S., Nishimura, H.: J. Geomag. Geolect. **16,** 72 (1964).
Hernandez, G., Turtle, J. P.: Planetary Space Sci. **17,** 675 (1969).
Herrmann, O.: Ann. Physik **25,** 143 (1936).
Herzenberg, A., Mandl, F.: Proc. Roy. Soc. (London) Ser. **A 270,** 48 (1962).
Hicks, G. T., Chubb, T. A.: J. Geophys. Res. **75,** 1290 (1970).
Holland, R. F.: Los Alamos Scientific. Lab., Rept. **LA-3783,** 1 (1967).
— Los Alamos Scientific Lab., Rept. **LA-DC-9468,** 1 (1968).
Holt, O., Lerfald, G. M.: Radio Sci. **2,** 1283 (1967).
Hundhausen, A. J., Ashbridge, J. R., Bame, S. J., Hilbert, H. E., Strong, I. B.: J. Geophys. Res. **72,** 87 (1967).
Hunten, D. M.: J. Atmospheric Terrest. Phys. **7,** 141 (1955).
— J. Atmospheric Terrest. Phys. **27,** 583 (1965).
— Can. J. Chem. **47,** 1875 (1969).
— McElroy, M. B.: Rev. Geophys. **4,** 303 (1966).
— — J. Geophys. Res. **73,** 2421 (1968).
Ivanchuk, V. I.: Sb. Rab. Mezhdunar Geofiz Godu (Collection of IGY papers) No **1,** 58 (1961).
Jobe, J. D., Sharpton, F. A., St. John, R. M.: J. Opt. Soc. Am. **57,** 106 (1967).
Johansen, O. E.: Planetary Space Sci. **14,** 217 (1966).
Jusick, A. F., Watson, C. E., Peterson, L. R., Green, A. E. S.: J. Geophys. Res. **72,** 3943 (1967).
Kamiyama, H.: Rept. Ionosph. Space Res. Japan **20,** 171 (1966).
Kishko, S. M., Kuchinka, M. I.: Opt. Spektroskopiya **6,** 378 (1959).
Koval, A. G., Koppe, V. T., Gritsyna, V. V., Fogel, Ya. M.: Geomagnetizm Aeronomiya **9,** 89 (Engl. transl.) (1969).
Lampe, F. W., Franklin, J. L., Field, F. H.: J. Am. Chem. Soc. **79,** 6129 (1957).
Landolt-Börnstein: Zahlenwerte und Funktionen, Berlin–Göttingen–Heidelberg: Springer 1952.
Langstroth, G. O.: Proc. Roy. Soc. (London) Ser. **A 146,** 166 (1934).
Lassettre, E. N., Glaser, F. M., Meyer, V. D., Skerbele, A.: J. Chem. Phys. **42,** 3429 (1965).
— Krasnow, M. E.: J. Chem. Phys. **40,** 1248 (1964).
— Silverman, S. M., Krasnow, M. E.: J. Chem. Phys. **40,** 1261 (1964).
Lytle, E. A., Hunten, D. M.: Can. J. Phys. **38,** 477 (1960).
Maeda, K., Aikin, A. C.: Report X-640-67-29, Goddard Space Flight Center 1967.
Mahadevan, P., Roach, F.: Nature **220,** 150 (1968).
Malville, J. M.: J. Atmospheric Terrest. Phys. **16,** 59 (1959).
McConkey, J. W., Latimer, J. D.: Proc. Phys. Soc. **86,** 463 (1965).
— Woolsey, J. M., Burns, D. J.: Planetary Space Sci. **15,** 1332 (1967).
McElroy, M. B.: Planetary Space Sci. **13,** 403 (1965).
McGowan, J. W., Clarke, E. M., Janson, H. P., Stebbings, R. F.: Phys. Rev. Letters **13,** 620 (1964).
McNamara, A. G.: Can. J. Phys. **47,** 1913 (1969).
Megill, L. R., Despain, A. M., Baker, D. J., Baker, K. D.: J. Geophys. Res. **75,** 4775 (1970).
Meyer, J. A., Setser, D. W., Stednan, D. H.: Astrophys. J. **157,** 1023 (1969).
Meyer, V. D., Lassettre, E. N.: J. Chem. Phys. **44,** 2535 (1966).
Mironov, A. V., Prokudina, V. S., Shefov, N. N.: Ann. Geophys, **14,** 364 (1958).
— — — In Spectral, Electrophotometrical, and Radar Researches of Aurorae and Airglow. No. 1, p. 20. Acad. Sci. Moscow 1959.

Mitra, S. K.: Sci. Cult. (Calcutta) **9,** 46 (1943).
Mukherjee, N. R.: Private comm. (1970).
Murcray, W.B.: Planetary Space Sci. **17**, 1429 (1969).
Nicholls, R.W.: J. Quant. Spectr. Radiative Transfer. **2**, 433 (1962).
Nishimura, H.: J. Phys. Soc. Japan **21**, 1018 (1966).
— J. Phys. Soc. Japan **24,** 130 (1968).
Noxon, J.F.: J. Geophys. Res. **75**, 1879 (1970).
Ogryzlo, E.A.: Can. J. Chem. **47**, 1871 (1969).
Omholt, A.: In The Airglow and the Aurorae. (Ed. E.B. Armstrong and A. Dalgarno) Pergamon Press 1956.
— J. Atomspheric Terrest. Phys. **10**, 320 (1957).
— Geofys. Publikasjoner **21**, No 1 (1959).
— Planetary Space Sci. **2**, 246 (1960).
— Phys. Norvegica **1**, 33 (1962).
— Ann. Geophys. **26** (1970).
— Harang, L.: J. Atmospheric Terrest. Phys. **7**, 247 (1955).
Parkinson, T.D., Zipf, E.C.: Planetary Space Sci. **18**, 895 (1970).
— Zipf, E.C. jr., Donahue, T.M.: Planetary Space Sci. **18**, 187 (1970).
Paulson, H. V., Shepherd, G. G.: J. Atmospheric Terrest Phys. **27,** 831 (1965).
Peterson, L.R., Green, A.E.S.: Proc. Phys. Soc. J. Phys. B. Ser. 2. **1**, 1131 (1968).
— Prasad, S.S., Green, A.E.S.: Can. J. Chem. **47**, 1774 (1969).
Peterson, V.L., Vanzandt, T.E.: Planetary Space Sci. **17**, 1725 (1969).
Philpot, J.L., Hughes, R.J.: Phys. Rev. **133 A,** 107 (1964).
Prasad, S.S., Green, A.E.S.: Trans. Am. Geophys. Union **51**, 368 (1970).
Rapp, D., Briglia, D.D.: J. Chem. Phys. **43**, 1480 (1965).
— Englander-Golden, P.G.: J. Chem. Phys. **43**, 1464 (1965).
Reasoner, D.L., Eather, R.H., O'Brien, B.J.: J. Geophys. Res. **73**, 4185 (1968).
Rees, M.H.: J. Atmospheric Terrest. Phys. **14**, 338 (1959).
— Stewart, A.T., Walker, J.C.G.: Planetary Space Sci. **17**, 1997 (1969).
— Walker, J.C.G.: In The Birkeland Symposium on Aurora and Magnetic Storms. (Ed. A. Egeland and J. Holtet. CNRS) 1967.
— — Dalgarno, A.: Planetary Space Sci. **15**, 1097 (1967).
Rothe, E.W., Marino, F.L., Neynaber, R.H., Trujillo, S.M.: Phys. Rev. **125**, 582 (1962).
Schram, B. L., de Heer, F. J., van der Wiel, M. J., Kistemaker, J.: Physica **31,** 94 (1965).
— Moustafa, H. R., Schutten, J., de Heer, F. J.: Physica **32,** 734 (1966).
Schulz, G.J.: Phys. Rev. **116**, 1141 (1959).
— Phys. Rev. **125**, 229 (1962).
— Phys. Rev. **135**, A 988 (1964).
— Dowell, J.T.: Phys. Rev. **128**, 174 (1962).
— Koons, H.C.: J. Chem. Phys. **44**, 1297 (1966).
Seaton, M.J.: J. Atmospheric Terrest. Phys. **4**, 295 (1954).
— In The Airglow and the Aurorae. (Ed. E.B. Armstrong and A. Dalgarno) Pergamon Press 1956.
— Phys. Rev. **113**, 814 (1959).
Shefov, N.N.: Planetary Space Sci. **5**, 75 (1961).
— Planetary Space Sci. **10**, 73 (1963).
Shemansky, D.F.: J. Chem. Phys. **51**, 689 (1969).
— Broadfoot, A.L.: Trans. Am. Geophys. Union **51**, 368 (1970).
— Carleton, N.P.: J. Chem. Phys. **51**, 682 (1969).
— Vallance Jones, A.: Planetary Space Sci. **16,** 115 (1968).

Sheridan, W.F., Oldenberg, O., Carleton, N.P.: Proc. Second Intern. Conf. Physics of Electronic and Atomic Collisions, Boulder Colo., p. 349 (1961).
Sheridan, J.R., Clark, K.C.: Phys. Rev. **140 A**, 1033 (1965).
Silverman, S.M., Lassettre, E.N.: J. Chem. Phys. **40**, 2922 (1964).
Simpson, F.R., McConkey, J.W.: Planetary Space Sci. **17**, 1941 (1969).
Smith, K., Henry, R.J.W., Burke, P.G.: Phys. Rev. **157**, 51 (1967).
Snelling, D.R., Bair, E.J.: J. Chem. Phys. **47**, 228 (1967).
Srivastava, B.N., Mirza, I.M.: Phys. Rev. **168**, 86 (1968).
Stauffer, A.D., McDowell, M.R.C.: Proc. Phys. Soc. **89**, 289 (1966).
Stebbings, R.F., Turner, B.R., Rutherford, J.A.: J. Geophys. Res. **71**, 771 (1966).
Stewart, D.T.: Proc. Phys. Soc. Ser. **A 69**, 437 (1956).
— Gabathuler, E.: Proc. Phys. Soc. **72**, 287 (1958).
Stoffregen, W.: Planetary Space Sci. **17**, 1927 (1969).
— J. Atmospheric Terrest. Phys. **32**, 171 (1970).
— Derblom, H.: Nature **185**, 28 (1960).
Stolarski, R.S., Dulock, V.A., Watson, C.E., Green, A.E. S.: J. Geophys. Res. **72**, 3953 (1967).
— Green, A.E.S.: J. Geophys. Res. **72**, 3967 (1967).
— Planetary Space Sci. **16**, 1265 (1968).
Strickland, D.J.: Trans. Am. Geophys. Union **51**, 368 (1970).
Stuhl, F., Welge, K.H.: Can. J. Chem. **47**, 1870 (1969).
Størmer, C.: The polar aurora. Oxford: Clarendon Press 1955.
Swider, W., Narcisi, R.S.: Planetary Space Sci. **18**, 379 (1970).
Swings, P.: In Atmospheres of the Earth and Planets. (Ed. G.P. Kuiper) 1948.
Takayanagi, K., Takahashi, T.: Rept. Ionosph. Space Res. Japan **20**, 357 (1966).
Tate, J.T., Smith, P.T.: Phys. Rev. **39**, 270 (1932).
Thieme, O.: Z. Physik **78**, 412 (1932).
Thompson, N., Williams, S.E.: Proc. Roy. Soc. Ser. **A 147**, 583 (1934).
Vallance Jones, A., Hunten, D.M.: Can. J. Phys. **38**, 458 (1960).
Vaysberg, O.L.: Polyarnyye siyaniya i svecheniye nochnogo neba, No **8**, 43 (1962).
Vegard, L., Kvifte, G., Omholt, A., Larsen, S.: Geophys. Publiskajoner **19**, no 3 (1955).
Wallace, L., Chamberlain, J.W.: Planetary Space Sci. **2**, 60 (1959).
— Hunten, D.M.: J. Geophys. Res. **73**, 4813 (1968).
— McElroy, M.B.: Planetary Space Sci. **14**, 677 (1966).
Wark, D.Q., Mercer, D.M.: Appl. Opt. **4**, 839 (1965).
Watson, C.E., Dulock, V.A., Stolarski, R.S., Croen, A.E.S.: J. Geophys. Res. **72**, 3961 (1967).
Williams, S.E.: Proc. Phys. Soc. Ser. **A 47**, 420 (1935).
Winters, H.F.: J. Chem. Phys. **44**, 1472 (1966).
Young, R.A., Black, G.: J. Chem. Phys. **44**, 3741 (1966).
— — J. Chem. Phys. **47**, 2311 (1967).
Zapesochnyi, I.P., Kishko, S.M.: Izv. Akad. Nauk SSSR **23**, 965 (1959).
— Skubenich, V.V.: Opt. Specktroskopiya **21**, 83 (1966).
Zare, R.N., Larsson, E.O. Berg, R.A.: J. Mol. Spectr. **15**, 117 (1965).
Zipf, E.C.: Bull. Am. Phys. Soc. **12**, 225 (1967).
— Can. J. Chem. **47**, 1863 (1969).

Chapter 6

Temperature Determinations from Auroral Emissions

6.1 Introduction

Determination of atmospheric temperatures from the auroral spectrum is based on measurements of: a) the thermal Doppler broadening of emission lines, and b) the rotational energy distribution of molecules, affecting the intensity distribution within a molecular emission band. Both methods implicitly assume that the excited atoms and molecules from which the radiation emerges are in thermal equilibrium with the surrounding atmosphere, or that the relation between the energies of the excited atoms and molecules can be unambiguously related to the atmospheric temperature.

The Doppler profile method is based on the use of atomic lines. The forbidden oxygen lines at 5577 and 6300/64 Å are used for these measurements. There are two reasons for this: First, these lines are among the most prominent in the most easily accessible wavelength region. Second, the excited, metastable states of atomic oxygen have lifetimes which are long enough to ensure that thermal equilibrium with the atmosphere is restored if it is disturbed by the excitation process. Furthermore, the self-absorption of these lines is negligible.

The rotational temperature method has been applied mainly to the first negative N_2^+ bands ($B^2\Sigma_u^+ - X^2\Sigma_g^+$). These bands are prominent in the violet part of the spectrum and easy to measure. Their rotational structure is such that the energy distribution among the rotational lines can be measured fairly accurately. In general, with excitation through electron impact on N_2, the rotational energy distribution of the excited molecules is believed to be close to thermal, so that it can be unambiguously related to the atmospheric temperature. However, for proton excitation this has been a problem of some concern, because it was thought that primary protons, with their relatively larger momentum, might significantly alter the rotational energy of the molecules, simultaneously with the ionization and excitation. This problem will be discussed in more detail in Sect. 6.3. Also for these bands self-absorption is unimportant, due to the low density of N_2^+ ions.

There have also been a few attempts to derive temperatures from the vibrational intensity distribution of molecular bands (cf. Chamberlain 1961, Hunten 1961). However, the energy difference between the vibrational levels in oxygen and nitrogen molecules is so large that higher vibrational levels in the ground state are not populated to any significant extent unless the temperature is above 1000 °K, which applies only to the highest altitudes. However, the intensity distribution among vibrational bands is most useful in elucidating excitation mechanisms (cf. Sect. 5.3.2).

In general, because of the rapid intensity and spatial variations occurring in aurora, temperatures inferred from auroral optical emissions are useful only when they are obtained over short time intervals. Because of this, and the superiority of photoelectric techniques, the earlier temperature measurements obtained by photographic techniques became almost immediately out-dated when photoelectric instruments were developed and used for optical temperature measurements.

6.2 Doppler Temperatures

The Doppler broadening of an emission line, due to the random thermal velocities of the atoms, is derived in standard texts. The intensity distribution in a line of frequency ν is given by

$$i(\nu) = \frac{I c}{\nu_o} \left(\frac{M}{2\pi k T} \right)^{\frac{1}{2}} \exp \left[- \frac{c^2 M}{2 k T} \left(\frac{\nu - \nu_o}{\nu_o} \right)^2 \right], \tag{6.1}$$

where I is the intensity of the line integrated over ν, ν_o is the center frequency, and the other symbols have their standard significance.

The half-width δ (distance between the point where the intensity is half of the maximum intensity) is given by

$$\delta = 2 (\log_e 2)^{\frac{1}{2}} \frac{\nu_0}{c} \left(\frac{2 k T}{M} \right)^{\frac{1}{2}} \tag{6.2}$$

The lifetimes against radiation of the $O(^1S)$ and $O(^1D)$ atoms are about 0.7 and 110 s respectively (cf. Sect. 5.2), and the collision frequency for an atom at heights between 100 and 170 km in the atmosphere is between 10^3 and $10\ s^{-1}$. Hence, at common auroral altitudes the $O(^1S)$ atoms will largely be restored to thermal conditions before radiation, even if serious distortion of velocities occurs during excitation; at altitudes lower than about 300 km the $O(^1D)$ atoms will be restored to thermal condition.

In addition, direct electron excitation of atomic oxygen will not affect the velocity of the atoms to any great extent, because of the small momenta of the electrons. Dissociative recombination of molecular oxygen, which probably is a minor source (cf. Sect. 5.1) gives a surplus energy of about 5.7 eV, less the excitation energy of 2 or 4 eV. This energy is to be divided among the two atoms, raising the velocity to a value which is 5 to 8 times the thermal one.

Because the thermal Doppler broadening of the lines is extremely small, high resolution optical instruments are necessary to obtain the desired resolution. The half-widths, in wave number, for the oxygen lines are of the order of magnitude 5×10^{-2} Kayser, or measured in wavelength 10^{-2} Å.

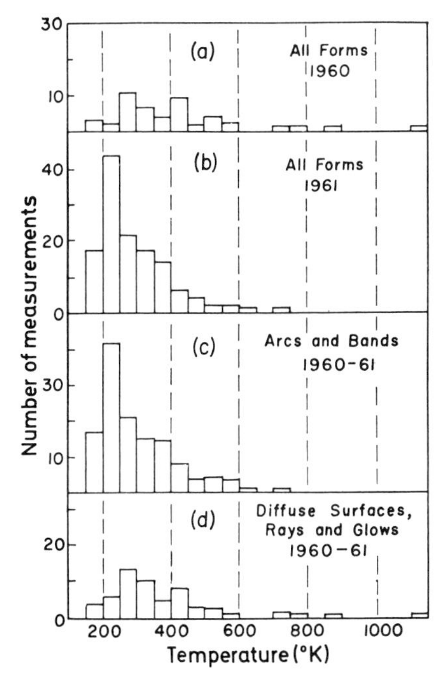

Fig. 6.1 Distribution of Doppler temperature measurements from the λ 5577 line (Turgeon and Shepherd 1962, courtesy Pergamon, Press.)

Not until Fabry-Perot interferometers with photoelectric recording were developed was it possible to obtain Doppler temperatures with reasonable resolution and time constant in the observations. Instruments were developed and used by several workers (Armstrong 1955, 1959, Chabbal 1958, Jacquinot 1960, Nilson and Shepherd 1961, Turgeon and Shepherd 1962). Because of the low light-gathering power of the Fabry-Perot instruments, even a moderately bright aurora could not be measured with scanning times less than 15 s if a reasonable accuracy (±50 °K) was to be obtained. Hilliard and Shepherd (1966) therefore developed a field-compensated Michelson interferometer for auroral and airglow temperature measurements.

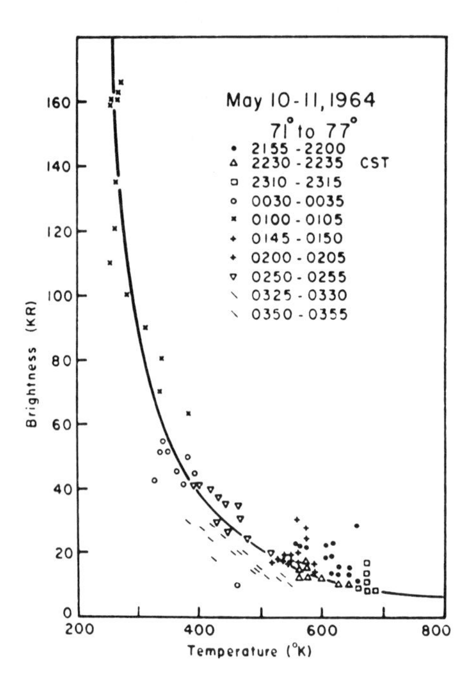

Fig. 6.2 Doppler temperature versus auroral brightness during one night (Hilliard and Shepherd 1966, courtesy Pergamon Press.)

The greatest success in measuring temperatures from Doppler profiles has undoubtedly been attained by Shepherd and his associates in the work referred to above. Nilson and Shepherd (1961) found Doppler temperatures from the λ 5577 line from 220 °K to 700 °K, with half of the data (12 of 26 measurements) falling between 400 and 500 °K. This work was continued by Turgeon and Shepherd (1962), who found a distribution of measured temperatures as shown in Fig. 6.1. From the λ 6300 line they found temperatures between 1000 °K and 1900 °K. Earlier, Wark (1960) and Mulyarchik (1959) had obtained temperatures of 730 °K and 3400 °K from this line, using photographic recording. In four cases Turgeon and Shepherd (1962) obtained vertical temperature scans through auroral forms, and were able to relate the temperatures to an approximate height in the atmosphere. Three of the cases gave the values 5.0, 4.2, and 4.0 °K km^{-1} for the temperature gradient between 100 and 160 km (with the lower border of the aurora assumed to be at 105 km). The fourth measurement, which the authors believe to be anomalous, gave 1.1 °K km^{-1}. Further, they found that the temperature determined by repeated measurements on one particular aurora could fluctuate 100 °K within 2—3 minutes. They also found, during one night, a significant decrease in the average temperature measured. Between 2200 and 0100 local time, the average temperature decreased from about 350 to 250 °K. This is in agreement with the decrease in auroral heights in the course of the night found by others (cf. Sect. 2.5).

Hilliard and Shepherd (1966) obtained an impressive series of λ 5577 Doppler temperatures by use of their Michelson interferometer. The most remarkable of their results is that there is often a systematic decrease in the Doppler temperature with increasing brightness over a certain time interval. The most striking example of this is shown in Fig. 6.2. The relation between temperature and intensity is not always as simple, and differs quantitatively from one case to another, but is still

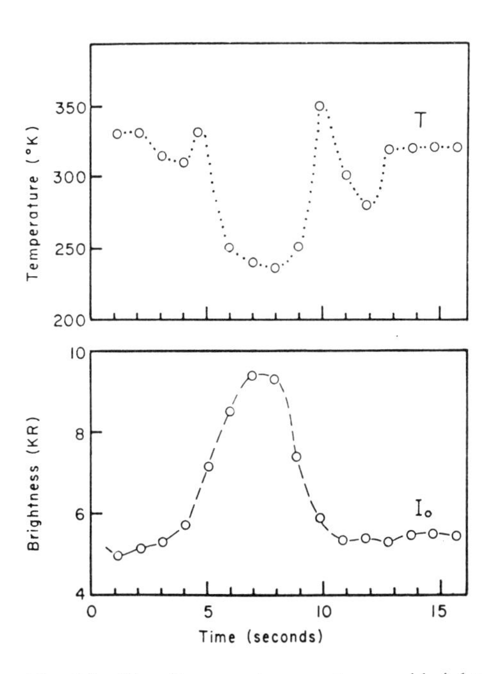

Fig. 6.3 Simultaneous temperature and brightness fluctuations in aurora (Hilliard and Shepherd 1966, courtesy Pergamon Press.)

striking. Even during short-period fluctuations in intensity, variations in temperature occur, as shown in Fig. 6.3.

The temperatures found by this method and their variation with apparent height seems reasonably consistent with current knowledge of the upper atmosphere. Hilliard and Shepherd (1966) interpreted their data in terms of a systematic variation of the height of the aurora with particle intensity. This would mean that the intensity modulation of an aurora is due to a significant extent to a modulation of the energy of the particles and not, or at least not wholly, due to a modulation of the particle flux. They found support for this view in statistical data on height measurements made by others (cf. Sect. 2.5). Adopting the luminosity profiles computed by Rees for exponential energy distribution (cf. Sect. 2.3), they found that when the characteristic (e-folding) energy increases from about 1.5 keV to 10 keV, the temperature deduced from the Doppler profile should decrease from about 500 °K to about 300 °K. With this interpretation, they found that the intensity variations and the associated variations in Doppler temperature could be almost entirely explained as due to variations in specific energy, except for the lowest energies (highest temperatures). The flux, however, varied considerably from one aurora to another.

There is no observational evidence that significant atmospheric heating occurs during auroras. This also holds for mid-latitude red arcs (Roble, Hays and Nagy 1970).

6.3 Rotational Temperatures

The rotational intensity distribution in molecular bands is ideally suited for temperature measurements, because the difference between the energy levels is such that a great number of rotational levels are populated, and the population distribution is very sensitive to the temperature of the gas. Also, excitation of molecules by electrons does not alter the rotational energy of the molecules, because the momentum of the electrons is small compared to the energy. For proton excitation there has been some concern about this, since it was thought that the heavier protons may change the rotation of the molecules. Fast protons ionize and excite directly as well as through secondary electrons, while slow protons also ionize and excite through charge exchange. However, laboratory evidence (cf. a detailed discussion by Vallance Jones and Hunten, 1960) as well as direct evidence from auroras (Gerard and Harang 1970) indicates strongly that protons do not change the rotational state of the molecules to any large extent.

The main problems in deriving useful temperatures from molecular bands are due to the structure of the bands, their intensity and to the spatial resolution and uncertainties in the determination of the height of the aurora. Structures of molecular bands are often such that there is considerable overlapping of rotational lines. Also, in particular with relatively weak bands, they may be badly blended by other auroral emissions. Sufficient intensity is also required in order to limit the time which is needed to determine the temperature with sufficient accuracy; otherwise the problems of spatial resolution and height location are seriously worsened.

For these reasons the (0—0) and (0—1) bands of the first negative N_2^+ band system, at 3914 and 4278 Å, are better suited than any other bands in the auroral spectrum. They constitute a sizeable fraction of the visible auroral light, and their structure is almost ideal for measurements. Although other band systems have been used for temperature measurements (cf. Chamberlain 1961, Hunten 1961, Mathews and Wallace 1961, Shemansky and Vallance Jones 1968, Ahmed 1969) most relevant data of this kind have been obtained through the first negative N_2^+ bands.

The general theory of the intensity distribution among the rotational lines in molecular bands is described in text books on molecular physics, and is not given here. For the first negative N_2^+ bands the intensity distribution in the socalled R-branch is given by:

$$I = \kappa K \exp[-hcBK(K+1)/kT], \tag{6.3}$$

where K is the rotational quantum number of the excited state, B is the rotational constant and κ is a constant given by molecular and general physical constants. A more accurate description of I includes a higher order term with $K^2(K+1)^2$ in the exponential, but this can for all practical purposes be neglected. Since the lifetime of the excited molecules is negligible, the thermal distribution is given by the neutral molecules in the ground state. The small change in internuclear distance is allowed for by some authors by using the B values of the ground state of N_2 rather than that of the $N^2\Sigma$ state of N_2^+. The difference, however, is very small and amounts to only about 4%, which is less than the usual error in the measurements. These bands contain two branches, the P- and R-branches. Within the P-branch there is considerable overlapping of rotational lines, while the R-branch shows a monotonically decreasing wavelength with increasing value of K, and with reasonable resolution. A scan of the λ 3914 bands is shown in Fig. 6.4.

Once the intensity as a function of K is known, the temperature may be determined, either from the maximum intensity point, which occurs for $K = K_1$ given by

$$K_1(2K_1+1) = \frac{kT}{hcB}, \tag{6.4}$$

or more reliably from fitting the observations to the relation

$$\log_e\left(\frac{I}{K}\right) = C - \frac{hcB}{kT}K(K+1). \tag{6.5}$$

$\log_e\left(\frac{I}{K}\right)$ versus $K(K+1)$ should give a straight line, the slope depending upon the temperature. If the observed rotational lines are not fully resolved, account must be taken of this, since the distance between lines increases with increasing K. This problem has been discussed in some detail by Shepherd and Hunten (1955).

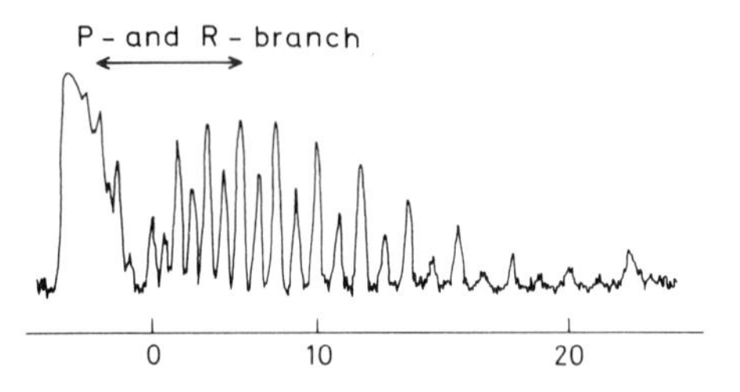

Fig. 6.4 Tracing from photographic spectrum of the (0—0) first negative N_2^+ band. Rotational quantum numbers for the R-branch are given below. (After Vallance Jones and Harrison 1955)

When $\log_e\left(\frac{I}{K}\right)$ is plotted against $K(K+1)$, one sometimes find that the points follow a slight curve rather than a straight line. This is to be expected if the radiation recorded originates from regions with different temperatures. Light from regions with high temperatures will contribute most to lines with high K, whereas the low temperature regions will contribute most to the intensity for low K's. Hence the deviation from a straight line gives an indication of the variability of the temperature in the emitting regions under observation, with the auroral intensity distribution as a weight function. Since an optical instrument on the ground always performs an integration of light along the line of sight, great care has to be taken to relate the observations to heights in the atmosphere.

A long series of rotational temperature measurements by photographic recording of only partly resolved spectra was obtained by Vegard and others (cf. Harang 1951, Hunten 1961, Chamberlain 1961). However, because of the long exposure times involved, the correlation with particular auroral forms and with height turned out to be very limited and in some cases non-existent. Thus, in his pioneer work Vegard was not able to find systematic temperature variations and erroneously concluded that the atmosphere was almost isothermal within the height interval most commonly covered by auroras.

The most successful use of photographic techniques appears to be the work by Vallance Jones and associates (Vallance Jones and Harrison 1955, Johanson and Vallance Jones 1962, Vallance Jones and Hunten 1960). Johanson and Vallance Jones (1962) were able to measure height variations of rotational temperatures in suitable auroral forms near the northern horizon. This was done by projecting the aurora on the spectrograph slit and assuming a height for the lower border. They found a temperature gradient of about $6\,^{\circ}\mathrm{K\,km^{-1}}$ in the 100 to 160 km region. Sunlit auroras showed temperatures between 800 and 1500 °K, placing these at great heights, in agreement with Størmer's results (1955). Also diffuse and pulsating auroras showed higher temperatures than did bands and arcs, but only by a few tens of degrees.

In the case of sunlit aurora, absorption of sunlight by N_2^+ ions produced by the primary particles contributes heavily to the excitation (cf. Sect. 5.6). The intensity distribution within and among the bands has to be calculated particularly from the excitation, taking the Fraunhofer lines into account. This has been discussed in detail by Vallance Jones and Hunten (1960).

More recent work has been based on photoelectric recording. Hunten *et al.* (1963) developed a "temperature photometer" which employed two narrow band filters centered at selected wavelengths within the λ 3914 band, such that the relative intensities recorded through the filters were sensitive to the rotational temperature. This permitted a temperature measurement within one second. A temperature profile was obtained, giving a gradient of about $6\,^{\circ}\mathrm{K\,km^{-1}}$. The same result was obtained by Måseide (1964), using a similar technique. He also observed a systematic temperature variation with time. The average temperature decreased by about one third during the first two hours after sunset, and then remained approximately constant. This result is similar to that of Turgeon and Shepherd (1962), reported in Sect. 6.2.

The most reliable rotational temperatures are those obtained by photoelectric spectrometers. A number of temperature measurements have been made by Shepherd and Hunten (1955), McEvan and Montalbetti (cf. Hunten 1961) and Brandy (1964, 1965). A temperature of about 200 °K at 90—100 km rising to about 500 °K at 150—160 km, giving a gradient of about 5—6 °K $\mathrm{km^{-1}}$, seems to be consistent with the results. Brandy's (1965) data indicate that the rotational temperature is above normal just before an auroral break-up event and below normal afterwards.

Temperatures have also been derived from the intensity distribution in the Meinel bands by Mathews and Wallace (1961). They found good agreement with data derived from the first negative bands.

Ahmed (1969) derived rotational temperatures from the Vegard-Kaplan bands, ranging from 900 to 2000 °K. These temperatures probably pertain to high altitudes. Shemansky (cf. Vallance Jones 1967) derived a temperature of 700 °K from the rotational distribution in the first positive bands.

6.4 Conclusions and Prospects

From the available optical temperature measurements the following important conclusions may be drawn:

a) The temperatures derived from auroral spectra give a temperature variation with height that is not quite consistent with current atmospheric models (CIRA 1965). The gradient of 6 °K, or slightly less, between 100 and 160 km indicated by auroral measurements is most consistent with the lowest solar activity. Even in this case, however, the CIRA model gives a slightly higher gradient.

Hence, there is an apparent disagreement between optical auroral temperatures and those obtained from the atmospheric density distribution, since it seems unrealistic to compare auroral temperatures with conditions of the lowest solar activity. Temperatures derived from artificial clouds of chemicals show temperatures consistent with the CIRA model (Harang 1969, 1970). In the latter type of measurement there is an unambiguous relation between temperature and height. This may not be so in the case of auroras. Whenever a distinct auroral form is present, there is usually also a weak glow or veil. Any observation from the ground will to some extent be contaminated by light originating at heights different from that of the distinct auroral form selected. Most of this light will probably originate at heights around 100—140 km, i.e. in regions with lower temperatures, and hence tend to make the temperatures observed at apparently greater heights too low. This difficulty may be overcome by making simultaneous observations inside and outside the distinct auroral form. There is little doubt that with a suitable technique auroras may be usefully employed for monitoring atmospheric temperatures.

b) Temperature variations strongly indicate significant and systematic variations in auroral heights, and consequently variations in the specific energy of the primary particles. Two kinds of variation are noted, both consistent with indications from other data: there is a variation in energy correlated with intensity variations during a particular auroral display, such that the intensity variations seem to a large degree due to variations in specific energy, and to a lesser degree due to variations in particle fluxes. There is also a decrease in temperature during the first part of the night, corresponding to a hardening of the energy spectrum of the primary particles.

c) There is no evidence of heating of the atmosphere during auroras.

d) Temperatures can now be measured optically from auroras with a time constant of seconds and with high accuracy (say 10% or better). This offers a possibility of monitoring the polar atmosphere and recording changes with solar cycle, solar activity, season and during the night. Systematic temperature measurements, together with simultaneous height measurements, offer a powerful and relatively inexpensive tool for such studies.

On the other hand, optical temperature measurements may offer a possibility to determine the height of more diffuse auroral forms, which are difficult to triangulate, in cases when the height is of interest for comparison with other data, for example radio reflection and absorption.

References

Ahmed, M.: J. Atmospheric Terrest. Phys. **31**, 1259 (1969).

Armstrong, E.B.: In The Airglow and the Aurora. (Ed. E.B. Armstrong and A. Dalgarno) Pergamon Press 1955.

— J. Atmospheric Terrest. Phys. **13**, 205 (1959).

Brandy, J.H.: Can. J. Phys. **42**, 1793 (1964).

— Can. J. Phys. **43**, 1697 (1965).

Chabbal, R.: Rev. Opt. **37**, 49 (1958).

Chamberlain, J.W.: Physics of the Aurora and Airglow. Academic Press 1961.

CIRA: Cospar International Reference Atmosphere. North-Holland Publ. Co. 1965.

Gerard, J.-C., Harang, O.: Phys. Norvegica, **4**, 217 (1970).

Harang, L.: The Aurorae. Chapman and Hall Ltd. 1951.

Harang, O.: In Atmospheric Emissions. (Ed. B.M. McCormac and A. Omholt) Van Nostrand Reinhold Co. 1969.

— Private communication 1970.

Hilliard, R.L., Shepherd, G.G.: Planetary Space Sci. **14**, 383 (1966).

Hunten, D.M.: Ann. Geophys. **17**, 249 (1961).

— Rawson, E.G., Walker, J.K.: Can. J. Phys. **41**, 258 (1963).

Jacquinot, P.: Rept. Progr. Phys. **23**, 267 (1960).

Johanson, A.E., Vallance Jones, A.: Can. J. Phys. **40**, 24 (1962).

Mathews, W.G., Wallace, L.: J. Atmospheric Terrest. Phys. **20**, 1 (1961).

Mulyarchik, T.M.: Spectral, Electrophotometrical, and Radar Researches of Aurora and Airglow. Acad. Sci. USSR., Moscow 1959.

Måseide, K.: In Studies of local morphology, structure and dynamics of aurora. (Ed. A. Omholt) Final report, Contract AF 61 (052—680. Blindern 1964.

Nilson, J.A., Shepherd, G.G.: Planetary Space Sci. **5**, 299 (1961).

Roble, R.G., Hays, P.B., Nagy, A.F.: Planetary Space Sci. **18**, 431 (1970).

Shemansky, D.E., Vallance Jones, A.: Planetary Space Sci. **16**, 1115 (1968).
Shepherd, G.G., Hunten, D.M.: J. Atmospheric Terrest. Phys. **6**, 328 (1955).
Størmer, C.: The Polar Aurora. Clarendon Press 1955.
Turgeon, E.C., Shepherd, G.G.: Planetary Space Sci. **9**, 295 (1962).
Vallance Jones, A.: In Aurora and Airglow. (Ed. B.M. McCormac) Reinhold Publ. Co. 1967.
— Harrison, A.W.: J. Atmospheric Terrest. Phys. **6**, 336 (1955).
— Hunten, D.M.: Can. J. Phys. **38**, 458 (1960).
Wark, D.Q.: Astrophys. J. **131**, 491 (1960).

Chapter 7

Pulsing Aurora

7.1 Introduction

In this chapter we shall discuss what is termed pulsing aurora. This includes pulsating, flaming, flickering and streaming forms. These forms of pulsing aurora are defined as follows in the International Auroral Atlas (1963):

p = Pulsing. Pulsing describes a condition of fairly rapid, often rhythmical fluctuations of brightness. The period of the fluctuation ranges from a fraction of a second to the order of minutes.

Four kinds of pulsing may be distinguished; they are described by the participles pulsating, flaming, flickering and streaming.

p_1 (Pulsating). This sub-class covers those conditions in which the phase of the variation of brightness is uniform throughout the form.

p_2 (Flaming). This sub-classification refers to a large area of the sky rather than to an individual form. The sky appears lit by surges of luminosity sweeping upwards towards the magnetic zenith. Very occasionally, flaming has been reported as sweeping downwards.

p_3 (Flickering). This is a condition of a large part a display which undergoes rapid, more or less irregular changes in brightness as if lit by flickering flames.

p_4 (Streaming). In this sub-class there is an irregular variation of brightness which progresses rapidly along the horizontal extent of homogeneous forms.

More or less rhythmic pulsations in the light output from auroral arcs, patches or diffuse surfaces have been noted for a long time as part of the auroral display. A description of the early observations, all of which were visual, is given by Størmer (1955). Since these pulsations in light intensity must reflect pulsations (modulations) in the incident stream of primary particles, the auroral pulsations must reflect similar variations in the magnetosphere. Hence the study of pulsating aurora may be a useful tool when studying instabilities and oscillations in the magnetospheric plasma.

The first attempt to describe the intensity variations in pulsating aurora appears to be that by Hassel in 1941, based on visual observations (cf. Størmer 1955). The fairly stationary geometry of the pulsating patches and arcs enabled Størmer to photograph them, and to measure their heights by triangulation. It appears from Størmer's data that the heights of pulsating arcs are not drastically different from those measured for quiet arcs (cf. Egeland and Omholt, 1966, 1967). There is evidence from Doppler temperature measurements, however, that the height of the aurora may vary during intensity fluctuations, it being lower during the periods of increased intensity (Hilliard and Shepherd 1966, cf. also Sect. 6.2). This conclusion was also reached by Parkinson, Zipf and Dick (1970) on the basis of a rocket observation of a pulsating aurora.

Visual pulsations are usually of low frequency, with periods of the order of magnitude of 10 seconds, but occasionally much more rapid pulsations are observed. From general experience, but not based on objective observations, the author is inclined to believe that rapid, visual pulsations are more commonly observed equatorwards of rather than in the auroral zone. However, there are no systematic observations yet to support this.

In recent years photoelectric techniques have made it possible to study pulsating aurora in more detail. Also it has been possible to study subvisual aurora, which often shows weaker and often faster pulsations than can be observed by the naked eye.

An early review of pulsating aurora was given by Campbell and Rees (1961). Other review papers are those by Shepherd and Pemberton (1968) and Omholt, Kvifte and Pettersen (1969).

7.2 Pulsating Aurora

7.2.1 Definition

Pulsating aurora is an aurora with approximately stable geometry which shows more or less rhythmic variations in intensity, all parts of the form varying approximately in phase. An idealized, stationary pulsating aurora may therefore most appropriately be defined by the equation

$$I(\boldsymbol{r},t)=I_S(\boldsymbol{r})\,I_T(t) \tag{7.1}$$

This means that the geometry of the aurora shows a stationary pattern described by the space function $I_S(\boldsymbol{r})$, and that the pulsations of the aurora are coherent at all points described by the time function $I_T(t)$. In this case the geometry can be studied, for example by photographic techniques, even with long exposure times, whereas the characteristics of the time variation can be studied photoelectrically at any point where the aurora has sufficient intensity to be recorded with high time resolution.

Observations, in particular by TV techniques (Davis 1969, Scourfield and Parsons 1969), show that Eq. (7.1) in many cases is not a very accurate description of pulsating aurora, and that the geometrical function $I_S(\boldsymbol{r})$ is varying irregularly, although less rapidly than the time function $I_T(t)$. Therefore, we cannot generally assume that the geometry is strictly stationary, and a more accurate description may be

$$I(\boldsymbol{r},t)=I_S(\boldsymbol{r},t)\,I_T(t)\,. \tag{7.2}$$

We assume here that $I_S(\boldsymbol{r},t)$ varies slowly with time compared to $I_T(t)$. The former function then describes the slowly varying geometry of the aurora, whereas the latter represents the coherent part of the variations, which we shall define as the pulsations.

When $F_S(\boldsymbol{r},\omega)$ and $F_T(\omega)$ describe the Fourier spectra of $I_S(\boldsymbol{r},t)$ and $I_T(t)$, the Fourier spectrum of $I(\boldsymbol{r},t)$ is described by

$$F(\boldsymbol{r},\omega)=\int_0^\infty F_S(\boldsymbol{r},y)\,F_T(\omega-y)\,\mathrm{d}y\,. \tag{7.3}$$

If $I_S(\boldsymbol{r},t)$ varies slowly with time, i.e. $F_S(\boldsymbol{r},\omega)$ has appreciable values for small values of ω only, and if the Fourier spectrum of $I_T(t)$ does not show narrow peaks, Eq. (7.3) may be written:

$$F(\boldsymbol{r},\omega)=F_T(\omega)\int_0^\infty F_S(\boldsymbol{r},y)\,\mathrm{d}y\,, \tag{7.4}$$

because $F_T(\omega-y)$ varies only little over that value of y for which $F_S(\boldsymbol{r},y)$ has any appreciable value. Hence, the power spectrum at higher frequencies is given by that of $I_T(t)$ in regions of ω where $F_S(\boldsymbol{r},t)$ does not show prominent, narrow peaks.

It may be reasonable to define a pulsating aurora as an aurora for which Eq. (7.4) is valid over a certain area and above a certain frequency, i.e. an aurora which shows coherent intensity variations over an area A_1 above a frequency ω_1.

Another point which must be considered is that a stable, pulsating auroral form may move across the sky. The function describing the pulsating aurora will obviously depend on the coordinate system of the observer. In a moving coordinate system the description will be different from that in a stationary one (if any such can be defined). Considering that the earth is a small object in space, there is no obvious reason why a coordinate system fixed on earth should be the most useful one for describing pulsating aurora. If we transform the function $I(\boldsymbol{r},t)$ to a moving coordinate system with respect to the earth, the functions $I_S(\boldsymbol{r},t)$ and $I_T(t)$ will also change. As is seen from the definition of pulsating aurora, the important achievement is to be able to separate the function $I(\boldsymbol{r},t)$ into $I_S(\boldsymbol{r},t)$ and $I_T(t)$ in such a manner that $I_S(\boldsymbol{r},t)$ shows as slow variations as possible. This means that the amplitude of the Fourier spectrum, $F_S(\boldsymbol{r},\omega)$ is as small as possible at higher frequencies. It seems therefore natural to seek a coordinate system for description which fulfills this in the best possible way. This can, in a very general but less precise way, be used to define the drift velocity of a pulsating aurora: the drift velocity is the velocity of that coordinate system which makes Eq. (7.4) valid to as low frequencies as possible and over as large an area as possible.

Only photoelectric equipment can be used for recording the dynamic characteristics of pulsating aurora. The time variations are best recorded by photoelectric photometers which are linear over a very wide dynamic range and have excellent sensitivity. The drift, however, is more difficult to measure by pointed photometers, which do not map the geometry, but only record the intensity at particular points. For this purpose TV camera techniques are superior (Cresswell and Davis 1966, Davis 1969), even if the linearity and dynamic range is quite limited compared to a simple photometer.

The pulsations preclude direct use of the simple technique used in ionospheric drift measurements: recording the amplitude at three different and suitably spaced points and measuring the phase shift. When a pulsating aurora has a drift velocity relative to the point of observation, a record by a stationary photometer will contain time variations which arise from the fact that the aurora is patchy and drifts by, in addition to the time variations due to pulsations of the aurora as seen in the drifting coordinate system. If the drift velocity is slow compared to the characteristic length of the patches, the drift effect will contribute to the low frequency part of the spectrum only. Hence, by some estimate of the scale size and drift velocity, it is possible to judge the validity of photometer records.

The simplest case appears when the true auroral pulsations do not contain low frequencies at all and the size of the patches and their drift velocity is such that they contribute to low frequencies only in the apparent time variations of the intensity. In this case the drift can be measured by studying the phase shift at low frequencies measured at three suitably spaced observing points (as is done in ionospheric drift measurements). In this case higher frequencies should show no phase shift between space points, being due to true pulsations as we have defined them. In certain frequency regions there may be a mixture of true pulsations and drift effects. Any method which implies a measure of the phase shift as a function of frequency may give some information on the drift velocity.

7.2.2 Observations

Quantitative information on the characteristics of optical pulsating aurora has been obtained by several authors (Heppner 1958, Murcray 1959, Campbell and Rees 1961, Iyengar and Shepherd 1961, Victor 1965, Paulson and Shepherd 1966a, b, Cresswell and Davis 1966, Johansen and Omholt 1966, Omholt and Pettersen 1967, Omholt and Berger 1967, Kvifte and Pettersen 1969, Davis 1969, cf. also Shepherd and Pemberton 1968). Since the observations have been made at different places and at arbitrary times, the data do not give a complete and systematic coverage, but only certain characteristics of single, pulsating auroras. Also some information on correlation with other phenomena and on the morphology of pulsating aurora is available (cf. Sects. 7.6 to 7.8).

Pulsating aurora occurs most often in the latter part of the night and is typically a post-break-up phenomenon in the auroral display. The pulsation frequency ranges from 0.01 to 10 Hz and the intensity is low, usually 1—2 kR or less. It never exceeds 10 kR as measured in λ 5577 or λ 3914. The most prominent and easily observed types of pulsations are fairly slow pulsating patches and arcs, with a pulsation time of several seconds. Examples of such low frequency pulsations are shown in Fig. 7.1. In some rare cases the pulsations may be more regular,

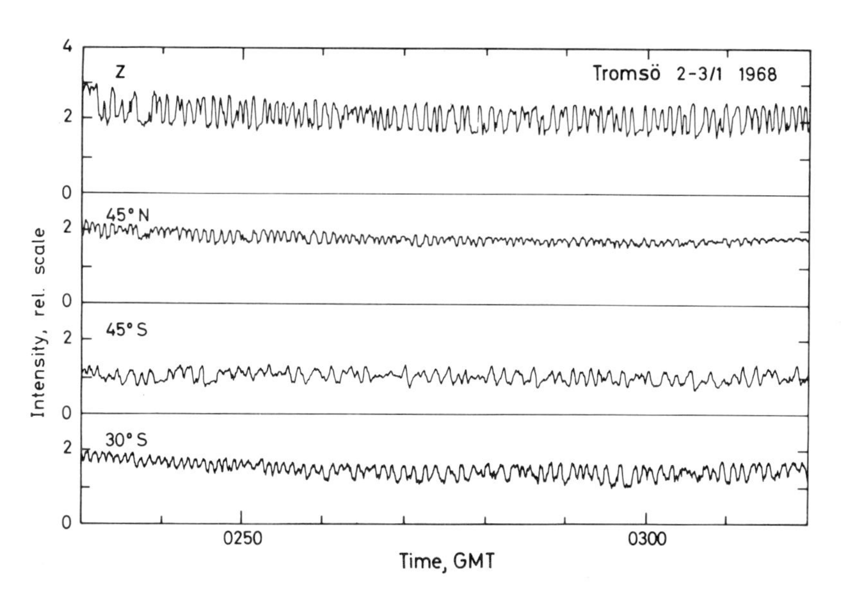

Fig. 7.1 Examples of low frequent pulsating aurora observed simultaneously in four different directions along the N—S meridian

even nearly sinusoidal (Johansen and Omholt 1966). The amplitude and Fourier spectrum may remain fairly constant for periods of several minutes, even when the pulsations are fairly irregular. Pulsations which have a sufficient intensity to be visual are most often of this type, at least this is so in and near the auroral zone. Fig. 7.2 shows the power spectrum of the pulsations shown in Fig. 7.1.

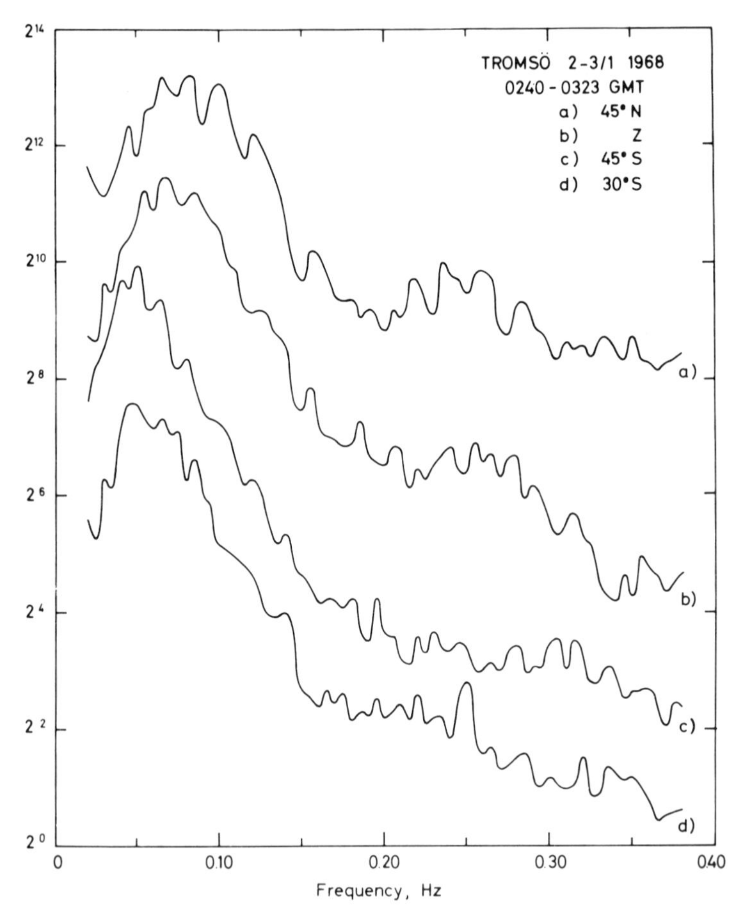

Fig. 7.2 Power spectra (spectral density) of the pulsations shown in Fig. 7.1. The spectra are on relative scale and are displaced in amplitude by a factor of 4 between each spectrum

At other times the pulsations are much more rapid. Most often (at least in the auroral zone) the rapid pulsations are subvisual, either occurring alone or superimposed on slower pulsations. Examples of such rapid pulsations are shown in Fig. 7.3 with an example of the high frequency part of the power spectrum in Fig. 7.4.

Johansen and Omholt (1966) also observed a particular, more rare type of pulsation with damped oscillations, characterized by one strong intensity peak followed by 5—8 pulses with decreasing amplitude. The time interval between the pulses was constant in each case, but ranged from one to three seconds.

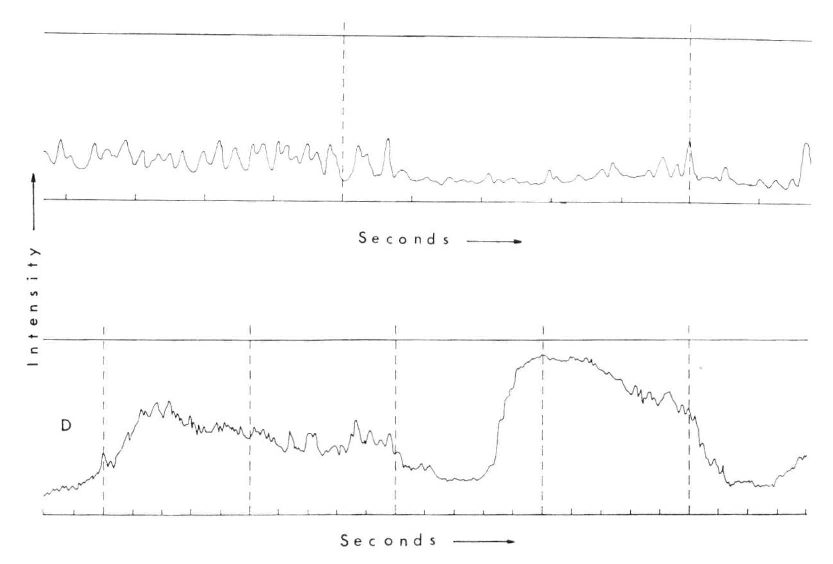

Fig. 7.3 High frequency auroral pulsations (from data by Omholt and Pettersen 1967)

Scourfield and Parsons (1969), using a TV technique, found that for subvisual pulsating patches the "switch on" and "switch off" of the luminosity was coherent over the patch. For stronger, visual patches, however, the "switch on" was coherent but not always the "switch off".

When the magnetic activity is high, a different type of pulsation is often observed in the morning hours (after 0600 local geomagnetic time). These are rapid, irregular pulsations, showing a striking similarity to auroral X-ray microbursts. They may occasionally have amplitudes exceeding the background by a factor of 3. Fig. 7.5 is an example of this type of pulsation. The power spectrum is very flat, and the most extended spectrum in Fig. 7.5 shows that significant power is present for frequencies as high as 2 Hz. Arnoldy (1970) has observed rapid fluctuations in the flux of precipitating electrons, with rocket techniques. Pulsations occurred in the 1—10 Hz frequency range and were associated with variations in pitch angle distribution.

Pulsating aurora may often cover a considerable part of the sky. Using 4 photometers, all equipped with N_2^+ filters at 4278 Å, Kvifte and Pettersen (cf. Omholt, Kvifte and Pettersen 1969) often observed pulsations simultaneously in the region from an elevation angle of 30°

Fig. 7.4 Power spectrum of the high frequency pulsations shown in the upper curve, Fig. 7.3 (Omholt and Pettersen 1967, courtesy Pergamon Press)

to the south to 45° to the north. Fig. 7.1 shows an example of this. The detailed correlation between the pulsations observed in the different directions is very low, possibly zero. However, the pulsations do have the same character on all channels. This is further demonstrated in Fig. 7.2, where the power spectra of these samples are shown. The power spectra are very similar, all having a principal peak in the frequency range 0.05—0.15 Hz. Power spectra of slowly pulsating aurora very often show broad peaks in this frequency range.

Looking in more detail, however, small significant differences are seen from one photometer record to another. As one moves from the most southern direction towards north, a small shift towards higher

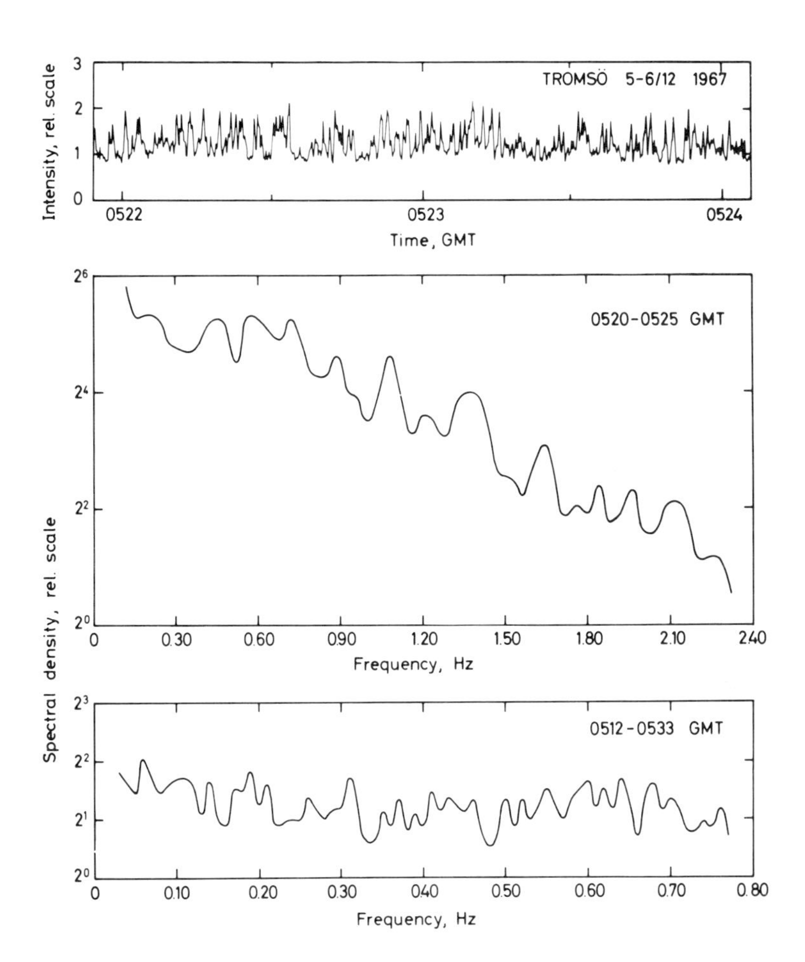

Fig. 7.5 Rapid, irregular fluctuations in auroral intensity during the morning hours. Intensity (above) and power spectrum (below)

frequencies is seen in the principal peak. Also, above this peak the spectrum seems to contain more high frequency power in the north and in the zenith than in the south.

This is contrary to variations of the average frequency with latitude over several displays. (Kvifte and Pettersen, private communication).

The average frequency of pulsating aurora decreases from about 0.06 Hz at 65° N geom. lat to about 0.03 Hz at 75° N geom. lat (Spitzbergen). Moreover, there is little or no variation in pulsation frequency at any one latitude with magnetic activity.

Observations by Shepherd and Pemberton (1968) and by the author (unpublished) indicate that the size of coherently pulsating spots is typically of the order of 5—10 km, although larger areas of coherent pulsations may occasionally be observed. This is particularly so with arcs. According to Cresswell and Davis (1966) each arc shows coherent pulsations at all points, whereas two neighbouring arcs may pulsate incoherently.

Gowell and Akasofu (1969) observed long-period pulsations in the brightness of the entire or a large part of the sky in the morning hours (0600—0700 UT). The period of pulsations was several minutes. These observations were made in the auroral zone by an all-sky camera, from an aircraft at a height of about 10 km. Similar, but rapid fluctuations in the λ 3914 intensity over the whole sky were reported by Murcray (1959).

Very few measurements have yet been made of the drift of the pulsating patches. Cresswell and Davis (1966) found eastward velocities of 1000 $m\,s^{-1}$ or less to be typical for fairly slowly pulsating aurora ($\simeq$a few seconds period).

Davis (1969) found evidence for conjugacy of pulsating aurora with a few seconds period. However, only one single example of this seems to exist.

Apparent pulsations in the $H\alpha$ and $H\beta$ intensities have been observed (Dzhordzhio 1962, Paulson and Shepherd 1966a). These observations were critically examined by Eather (1968), who also presented some new measurements. He concluded that the earlier examples of pulsations in $H\beta$ were most likely due to contamination by other radiation, since the interference filters used in such studies have a finite band-width and wings in the transmission curves, almost certainly admitting a small fraction of other auroral light. This was also suspected by Shepherd and Pemberton (1968).

Careful simultaneous measurements of electron-excited emissions and $H\beta$ emission during three pulsating auroras, with 5—15 s periods, led Eather (1968) to the conclusion that there were no detectable H β pulsations, whereas in an event with 50—200 s pulsating period both H β and the λ 5577 [OI] line were modulated to the same extent. The extimated contribution from protons to the total excitation of the aurora was about 15—20%. Another phenomenon, short-lived bursts of $H\beta$ emission, seems to be real, although rare (cf. Sect. 7.4).

7.3 Flaming Aurora

Flaming aurora consists of waves of light, with an arc-like geometry, apparently travelling upwards from the 100 km region with high velocity. This is a fairly rare phenomenon, typically post-break-up. In the author's opinion it occurs relatively more often in cis-auroral than auroral regions. According to Størmer (1955) flaming aurora is often followed by a corona. Carleton (cf. Omholt 1963) measured the speed of these "waves" with a photoelectric technique, and found velocities of the order of 1000 km s^{-1}. This may suggest the following interpretation: If a great number of electrons with different energies are released simultaneously from the same place in the magnetosphere, and travel along the field lines toward the polar regions, the fast electrons will arrive first and also penetrate to the lowest heights. The slower electrons will arrive in succession according to speed, and will be stopped (and produce aurora) at successively increasing altitude, the results being an emission region travelling upwards as observed. From his measurements Carleton inferred a distance of about 2500 km to the source. If this explanation is correct, it suggests that trapped particles are released by some disturbance operating for a fraction of a second. The light emission from any particular place lasts only a few tenths of a second.

Creswell (1968a), using TV technique, reported upward travelling waves having velocities of about 80 km s^{-1}, while Scourfield and Parsons (1969) found velocities between 75 and 130 km s^{-1}. They find that the source for the electrons in this case must be placed at a distance along the field line ranging from 2.5 to 4 earth radii.

7.4 Flickering Aurora

This is a fairly rare condition which has been observed in auroral and cis-auroral regions. Its unique characteristic is that a large part of the aurora undergoes rapid, irregular or regular changes in brightness. The only reasonably comprehensive quantitative study of this phenomenon is that of Beach *et al.* (1968), by visual and by image-orthicon television observations. They observed flickering aurora occurring just before and during the auroral break-up. The frequency was relatively fixed, 10 ± 3 Hz, and the flickering features were mainly in the form of interconnected spots of diameter roughly 1—5 km. Flickering auroras were observed at various locations ranging in geomagnetic latitude from 63° to 68°, in the northern as well as the southern hemisphere. All observations were made in the few hours preceding local geomagnetic midnight. Without exception, the flickering was observed to occur within bright homogeneous arcs or within bands which showed rapid internal motions (>5 km s^{-1}).

On one occasion a zenith photometer recorded unusually short-lived bursts of $H\beta$ emission with intensities up to 650 R within the flickering form. From riometer observations they concluded that there was also precipitation of large fluxes of electrons with energies of 30 keV and above. Short-lived bursts of $H\beta$ emission have also been observed by Måseide (private communication). They must be rare events. Evans (1967) has observed periodic 10 Hz fluctuations in precipitated electrons in the 1—120 keV energy range. In that case there was also a strong monoenergetic component of 6 keV in the electron flux. Paulson and Shepherd (1966b) observed similar pulsations in active aurora, also with periods of about 10 Hz (± 3 Hz).

Conjugacy of flickering aurora was observed by Beach et al. (1968), recording simultaneously bright, homogeneous bands with flickering from central Alaska and south New Zealand. Although observations were obtained during a very short time interval only (about one minute) it indicates that flickering is a conjugate phenomenon.

7.5 Streaming Aurora and Horizontal Waves

Streaming aurora is a brightening of a part of an auroral arc or band, travelling horizontally along the form of the wave, without changing the geometry of the form. This phenomenon is very common during and near the break-up phase of an auroral display.

Another particular form of horizontal waves was detected by Cresswell and Belon (1966) and is further described by Cresswell (1968b). These are weak east-west aligned, arc-like forms. The entire form travels horizontally and equatorwards, emerging from auroral arcs or glow, with typical velocities of 50—300 km s^{-1}. They occur in the post-midnight hours, and only when the preceding daylight hours, and often the evening hours, were magnetically quiet, and are always preceded by short-lived (1—2 hours) polar substorms. They occur with a frequency of about one per second, travel a distance of 150 km or more, have a width of 30—120 km and an intensity of less than 1 kR in λ 4278. The spectra sometimes show very weak $H\beta$ emission, but generally indicate electron excitation. Cresswell (1968b) associates these waves with hydromagnetic processes occurring near the equatorial plane. The hydromagnetic velocity normal to the field in this region gives velocities of the order of 50 km s^{-1} when projected on to the auroral regions along the field lines.

7.6 Correlation with Pulsations in the Magnetic Field and Telluric Currents

Magnetic disturbances and telluric currents are strongly linked to auroral displays. It is therefore of interest to investigate whether auroral pulsations are directly correlated with pulsations observed on magnetic records or on records of telluric currents.

Campbell and Rees (1961) studied the correlation of auroral pulsations having 6 to 10 s period with magnetic micropulsations (Alaska 64° N, geom. lat.). They found that the general amplitude variations of the auroral pulsations were quite similar to those of the magnetic field micropulsations and that the intensity of the two phenomena had a correlation coefficient of about 0.6. It was possible to find peak to peak correspondence between light pulsations and magnetic micropulsations for about 60 per cent of the time that auroral pulsations appeared. Omholt and Berger (1967) made attempts to correlate frequencies of optical and magnetic pulsations occurring simultaneously, but failed to find such correlation (Tromsø, 67° N. geom. lat.) Berger (1963) on one occasion found an excellent correlation between a strong event of slow auroral pulsations and giant magnetic pulsations. Victor (1965) made simultaneous observations of auroral and geomagnetic fluctuations at Byrd station (70° S geom. lat.). A one-to-one correspondence was often found between the peaks of auroral fluctuations and peaks of the geomagnetic micropulsations. This correspondence was most likely to be observed at times of regular auroral pulsations, but was also observed to continue for long time intervals which included a number of regular auroral pulsations separated by sequences of irregular fluctuations.

Paulson, Shepherd and Graystone (1967) compared geomagnetic and telluric-current fluctuations with auroral pulsations (Saskatoon, 60° N geom. lat.) They also found correlation between the two sets of phenomena, and made careful computations to obtain a quantitative measure of this correlation. A positive cross-correlation was found, and the spectral density of the two phenomena was similar in shape. The maximum of the cross-correlation function for the auroral and telluric-current fluctuations was displaced 1.6 s corresponding to a phase lag for the auroral fluctuations.

There is at present no clear picture of the relation between auroral pulsations on the one hand and magnetic and telluric-current pulsations on the other. There is apparently a limited correlation, but the relation may differ from time to time and there may be differences in the detailed appearance. This is to be expected from the very nature of the two phenomena. Even if auroral pulsations were directly accompanied by increased current density in that part of the ionosphere where the pulsations occurred, there is likely to be a considerable difference in the phenomena as observed from the ground. The two types of observations, optical and magnetic, both include integration of effects from a large part of the ionosphere, but with a different weight distribution function. A one-to-one correspondence between peaks is therefore to be expected only when the aurora pulsates coherently over a large area. If there are pulsations over large areas in the sky, incoherent but with the same spectral distribution, a peak-to-peak correspondence in the records is not to be expected, but the power distribution in the spectrum should be similar for the two phenomena.

7.7 Correlation with Pulsations in X-Rays

It is also well known that auroral X-rays show pulsations (cf. Chapt. 9) Whereas auroral light is produced mainly by electrons in the 1—10 keV region, the X-rays observed from balloons to date are confined to higher energies, from about 20 keV and upwards. Hence comparison of auroral pulsations and X-ray pulsations is a comparison of pulsations in the fluxes of lower and higher energy electrons. Such comparisons, together with studies of the frequency spectrum of the pulsations, may eventually give a clue to the modulation mechanisms responsible for the pulsations. However, for direct comparison between auroral luminosity and auroral X-rays, geometrical factors obscure the relation.

There are a number of observations showing pulsations and so-called microbursts in auroral X-rays. When looking at the records of microbursts and of rapid auroral pulsations, one gets the impression that the two phenomena are very similar. Kvifte and Pettersen (private communication) often observe pulsations in auroral light similar to X-ray microbursts in the morning hours (Tromsø, 67° N geom. lat.) However, these microbursts, which occur mainly in the morning hours, may well have a different origin than the most common types of auroral pulsations. Lampton (1967) observed that the low-energy particles (13 keV and 7 keV), which carried a large part of the particle flux, did not take part in the pulsations.

Simultaneous pulsations in aurora and X-rays were observed by Rosenberg *et al.* (1967). This was a rather typical event of auroral pulsations. In this case both high- and low-energy particles took part in the pulsations. It is estimated (Rosenberg *et al.* 1971) that 25—50 per cent of the modulated flux was due to particles with energy greater than 25 keV, the rest being due to particles of lower energy.

McPherron *et al.* (1968) occasionally found a similarity between the Fourier spectra of X-ray pulsations and geomagnetic micropulsations, but perfect agreement between spectral details did not occur. The micropulsation events usually lasted much longer than the X-ray pulsation events. Milton *et al.* (1967) in some cases found direct correspondence between X-ray microbursts and impulsive micropulsations.

7.8 Morphology of Pulsating Aurora

To get a clear picture of pulsating aurora as a phenomenon, it is necessary to know not only the characteristics of individual pulsating auroras, but also their distribution in a suitably oriented coordinate system and their relation to the substorm. We have in the past and probably in the near future must still rely on single-station observations of pulsating aurora; hence a statistical approach is necessary.

Akasofu (1964, 1968) noted from his studies of auroral substorms that drifting patches, which according to common experience often are pulsating, occur mainly during the recovery phase of the substorm and on the morning side. With an optical all-sky camera, none but the very slowest pulsations can be traced as such, but pulsating aurora is usually recognizable to an experienced observer by its geometry. Therefore, it has also been possible to do some morphological work on pulsating aurora by use of all-sky cameras (cf. Heppner 1954, 1958).

Up to the work by Kvifte and Pettersen (1969), few papers gave more than a qualitative statement on the morphological picture of pulsating aurora. However, a general agreement seemed to exist between the various workers on the dependence of pulsating aurora on geomagnetic latitude and local time. Størmer's data from the years 1911 to 1944 (cf. Egeland and Omholt 1966, 1967) indicate that pulsating surfaces are common at least in the region 59—67° N geom. lat., while pulsating arcs are very rare north of 64° N geom. lat. Heppner (1954, 1958) found similar results, reporting that pulsating surfaces and arcs are rare outside a belt between 60 and 67° N geom. lat. He also observed that the pulsations occurred later in the night in the northern than in the southern part of the belt. Cresswell and Davis (1966) confirmed the latitude dependence found by Heppner. They noted that the pulsations usually take place at the equatorward boundary of the auroral display.

Most authors have found that the occurrence of pulsations depends strongly on local time. Heppner (1954, 1958), Victor (1965), Cresswell and Davis (1966) and Omholt and Berger (1967) all report that pulsations occur almost exclusively after local magnetic midnight. Størmer (1955) observed that pulsating arcs sometimes occurred in the evening when a great auroral event had covered most of the sky for a night or two. Campbell and Rees (1961) found that the greatest pulsation amplitudes occurred after local midnight, when a constant plateau was reached. Just before dawn the amplitudes again seemed to increase. At this time, shorter periods than the most common ones of 5—20 s were seen. Omholt and Berger (1967) found a continuous decrease in amplitude after magnetic midnight, for pulsations with periods above 10 s. Paulson and Shepherd (1966a, b) found that rapid fluctuations in the auroral light occurred at all times and in almost all quiet forms. These were often very rapid and with small relative amplitude. Their observations were made at low latitudes.

Kvifte and Pettersen (1969) made a comprehensive study of the morphology of pulsating aurora at selected latitudes (cf. also Omholt, Kvifte and Pettersen 1969). They measured with photoelectric photometers in fixed directions from Tromsø (67° N geom. lat.), covering a latitude range of about 3° (cf. Fig. 7.1), and also used photometer data from Spitzbergen (75° N geom. lat.) obtained by Brekke (1969).

The instruments at Tromsø recorded pulsations in the frequency range 0—20 Hz. Pulsating aurora was observed on 27 clear nights during the winter of 1967/1968. The occurrence showed a definite dependence on magnetic activity, as defined by the 24-hour sum of local K-indices centered around midnight of the night of observation. The results for high magnetic activity, defined by a sum K value greater than 20, are summarized in Fig. 7.6a, and for low magnetic activity in Fig. 7.6b. All data from Spitzbergen are included in both figures, because data were scarce and no significant dependence on magnetic activity was found. The numbers give the percentage of all time during these nights that pulsating aurora in some form occurred. As is seen, the maximum of occurence at 67° N. geom. lat. is around 0600 geom. time, and at about 1000 geom. time at 75°. At Spitzbergen the maximum has a much lower value than in Tromsø. Later data, covering the evening sector from about 1900, indicate that pulsations also occur in the evening, independently of magnetic activity, but with low probability and small pulsation amplitude, such that visual pulsations are rare, although they do exist (Kvifte and Pettersen, private communication).

Figs. 7.6 a and b clearly demonstrate, despite the limited coverage in latitude, that pulsating aurora occurs in a more or less spiral-shaped region or part of an oval which expands equatorwards when the magnetic activity increases. Comparison with the oval for occurrence of discrete auroras (Feldstein 1966, Feldstein and Starkov 1967, Hartz and Brice 1967) shows that the maximum of the pulsating aurora is just south of or in the southern part of the oval.

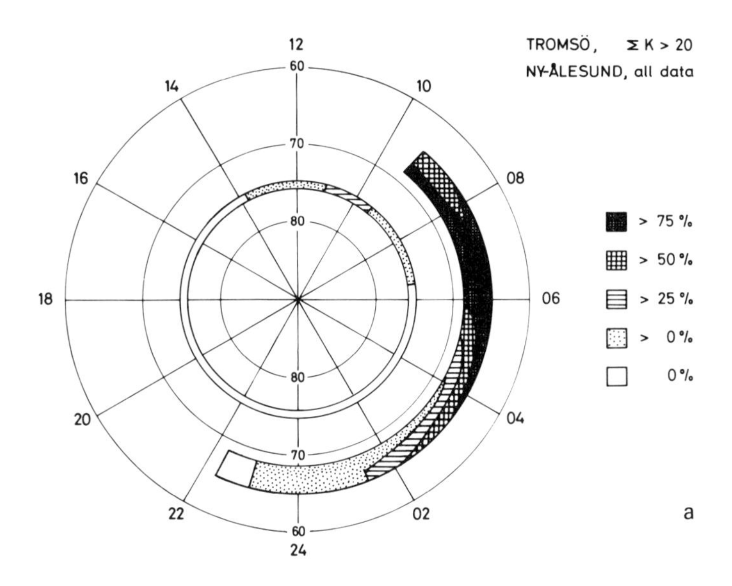

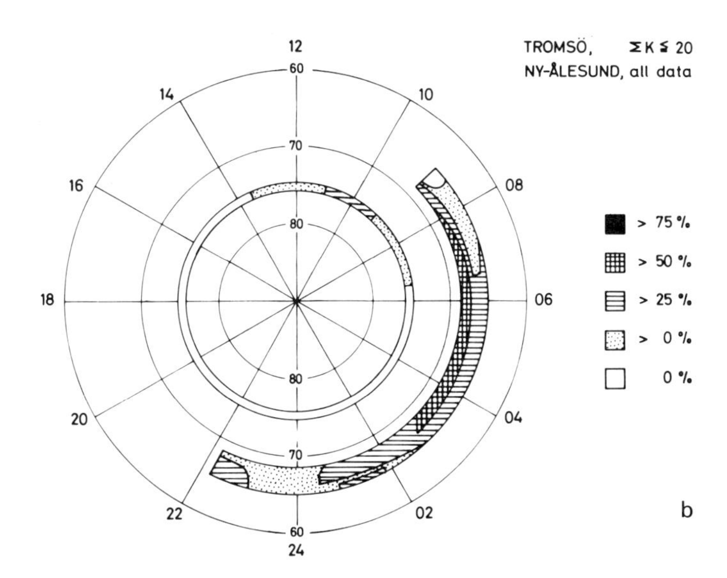

Fig. 7.6 Occurrence of auroral pulsations in the frequency range 0.01—20 Hz during high (a) and low (b) magnetic activity. Magnetic coordinates.

There is general agreement with earlier work. The slight differences which occur are attributable to a tendency to a change to higher pulsation frequencies in the morning hours, to which the instruments used by earlier workers were not so sensitive.

These results agree, generally, with those of Cresswell (1970), who investigated the morphology of aurora by combining several years of visual, photometer, television and all-sky camera observations from College, Alaska. The pattern of occurrence is similar to that found by Kvifte and Pettersen (1969). Further, Cresswell found that pulsating auroras may develop over a station in several ways, the most common being: (i) in the wake of a poleward expansion, (ii) within a diffuse envelope of aurora that spreads equatorwards, and (iii) as eastwards drifting patches that form equatorwards of other auroras. Pulsating aurora in the form of patches was most common in the morning hours. Starkov and Rolgodin (1970) have found a similar pattern.

The average amplitude of the pulsations in the λ 4278 N_2^+ band observed by Kvifte and Pettersen (1969) was about 400 R for the most disturbed nights ($\Sigma K > 20$) and about 100 R for less disturbed nights. In the latter case it decreased monotonously from 200 R at 2200 geom. time to about 30 R at 0400 geom. time, whereas during disturbed nights it increased abruptly during the first couple of hours after onset and remained approximately constant thereafter. The average modulation (pulsation amplitude relative to the constant background) was 10 to 30% for the most disturbed and 10 to 15% for less disturbed nights. During the most disturbed nights ($\Sigma K > 20$) the modulation varied in the course of the night, from about 10% at 22 geom. time to about 30% at 06 geom. time. The average frequency of the pulsations depends on geomagnetic latitude, being lower at high latitudes than at lower ones (cf. Sect. 7.2.2).

Parks *et al.* (1968) have reviewed and studied the occurrence of X-ray pulsations and bursts as a function of local time, and in a subsequent paper Coroniti *et al.* (1968) studied their relation to the magnetic substorm. They find that the various temporal features of auroral-zone energetic electron precipitation occur primarily during times when an auroral substorm is in progress, but that there are distinct temporal characteristics for each local time. In a paper by McPherron *et al.* (1968) the pulsation characteristics are summarized in a polar diagram. This shows that the night side, where pulsations in the light may be observed, is dominated by noise bursts and impulses, rather than quasi-sinusoidal pulsations. The latter occur mainly in the midday and evening.

The study of pulsation morphology is still at its beginning; much more work is required to make it a useful tool for studies of magnetospheric phenomena.

References

Akasofu, S.-I.: Planetary Space Sci. **12**, 273 (1964).
— Polar and Magnetic Substorms. Reidel Publ. Co. 1968.
Arnoldy, R. L.: J. Geophys. Res. **75**, 228 (1970).
Beach, R., Creswell, G. R., Davis, T. N., Hallinan, T. I., Sweet, L. R.: Planetary Space Sci, **16**, 1525 (1968).
Berger, S.: Planetary Space Sci. **11**, 867 (1963).
Brekke, A.: Thesis. Univ. of Oslo 1969.
Campbell, W. W., Rees, M. H.: J. Geophys. Res. **66**, 41 (1961).
Coroniti, F. V., Kennel, C. F.: J. Geophys. Res. **75**, 1279 (1970).
— McPherron, R. L. Parks, G. K.: J. Geophys. Res. **73**, 1715 (1968).
Cresswell, G. R.: Rep. UAGR-206, Geophysical Institute, Univ. of Alaska (1968a).
— Planetary Space Sci. **16**, 1453 (1968b).
— On pulsating aurora morphology. Report 1970.
— Belon, A. E.: Planetary Space Sci. **14**, 299 (1966).
— Davis, T. N.: J. Geophys. Res. **71**, 3155 (1966).
Davis, T. N.: In Atmospheric Emissions. (Ed. B. M. McCormac and A. Omholt) Van Nostrand Reinhold Co. 1969.
Dzhordzhio, N. V.: Coruscation regularities in aurora. Akad. Nauk. USSR (NASA Tech. Trans. TTF-20) 1962.
Eather, R. H.: The Birkeland Symposium on Aurora and Magnetic Storms. (Ed. J. Holtet and A. Egeland) p. 111 (1968).
Egeland, A., Omholt, A.: Geofys. Publikasjoner **26**, No 6 (1966).
— — In Aurora and Airglow. (Ed. B. M. McCormac) p. 143, Reinhold Publ. Co. 1967.
Evans, D. S.: J. Geophys. Res. **72**, 4281 (1967).
Feldstein, Y. I.: Planetary Space Sci. **14**, 121 (1966).
— Starkov, G. V.: Planetary Space Sci. **15**, 209 (1967).
Gowell, R. W., Akasofu, S.-I.: Planetary Space Sci. **17**, 289 (1969).
Hartz, T. R., Brice, N. M.: Planetary Space Sci. **15**, 301 (1967).
Heppner, J. P.: J. Geophys. Res. **59**, 329 (1954).
— Report No. D. R. 135, Defence Research Board. Canada 1958.
Hilliard, R. L., Shepherd, G. G.: Planetary Space Sci. **14**, 383 (1966).
International Auroral Atlas: Edinburgh University Press 1963.
Iyengar, R. S., Shepherd, G. G.: Can. J. Phys. **39**, 1911 (1961).
Johansen, O. E., Omholt, A.: Planetary Space Sci. **14**, 207 (1966).
Kvifte, G. J., Pettersen, H.: Planetary Space Sci. **17**, 1599 (1969).
Lampton, M.: J. Geophys. Res. **72**, 5817 (1967).
McPherron, R. L., Parks, G. K., Coroniti, F. V., Ward, S. H.: J. Geophys. Res. **73**, 1697 (1968).
Milton, D. W., McPherron, R. L., Anderson, K. A., Ward, S. H.: J. Geophys. Res. **72**, 414 (1967).
Murcray, W. B.: J. Geophys. Res. **64**, 955 (1959).
Omholt, A.: Planetary Space Sci. **10**, 247 (1963).
— Berger, S.: Planetary Space Sci. **15**, 1075 (1967).
— Kvifte, G., Pettersen, H.: In Atmospheric Emissions. (Ed. B. M. McCormac and A. Omholt) Van Nostrand Reinhold Co. 1969.
— Pettersen, H.: Planetary Space Sci. **15**, 347 (1967).
Parks, G. K., Coroniti, F. V., McPherron, R. L., Anderson, K. A.: J. Geophys. Res. **73**, 1685 (1968).

Parkinson, T.D., Zipf, E.C., Dick, K.A.: J. Geophys. Res. **75**, 1334 (1970).
Paulson, K.V., Shepherd, G.G.: Can. J. Phys. **44**, 837 (1966a).
— — Can. J. Phys. **44**, 921 (1966b).
— — Graystone, P.: Can. J. Phys. **45**, 2813 (1967).
Rosenberg, T.J., Bjordal, J., Kvifte, G.J.: J. Geophys. Res. **72**, 3504 (1967).
— — Trefall, H., Kvifte, G. J., Omholt, A., Egeland, A.: J. Geophys Res. **76**, 122 (1971).
Scourfield, M. W. J., Parsons, N. R.: Planetary Space Sci. **17**, 1141 (1969).
Shepherd, G.G., Pemberton, E.V.: Radio Sci. **3**, 650 (1968).
Starkov, G.V., Rolgodin, V.K.: Geomagnetizm i Aeronomiya **10**, 97 (Russ.) (1970).
Størmer, C.: The Polar Aurora, Oxford: Clarendon Press 1955.
Victor, L.J.: J. Geophys. Res. **70**, 3123 (1965).

Chapter 8

Optical Aurora and Radio Observations

8.1 Introduction

The detailed relationship between the optical emission from auroras and the enhanced rate of ionization associated with it was discussed in Sect. 2.2. This chapter is devoted to the association between the optical aurora and the effects of the enhanced ionization on radio wave propagation. There are several important and distinctly different radio phenomena which are associated with aurora: local absorption of radio waves; reflection from an extraordinary E-layer formed by diffuse, extended auroras; scattering of radio waves that are not absorbed; VLF radio emissions and radio noise, and the extended, more rare polar cap absorption events. Auroras giving rise to backscatter of radio waves that are not absorbed are often called radio auroras.

We shall not discuss the radio phenomena at any length or in detail —this will be done in a coming volume of *Physics and Chemistry in Space*—but rather concentrate on the direct association of these with optical aurora. Since radio observations can be performed regardless of weather conditions, much valuable work has been done with this technique on general auroral morphology and the relationship between the aurora and magnetospheric phenomena.

8.2 Radio Absorption

It is a well-known phenomenon that absorption of radio waves in the ionosphere accompanies optical aurora (see e.g. Hultqvist 1965, 1969, Little 1967, Hargreaves 1969). The question to be discussed here is how this correlation appears and how it may be explained, without going into details of the radio absorption mechanism or the technique used for the measurements. A comprehensive review of auroral absorption of high-frequency waves is given by Hargreaves (1969). The existence of, at least, a statistical relationship between auroral luminosity and absorption of radio waves has been established by several workers (Heppner *et al.*

1952, Little and Leinbach 1958, Campbell and Leinbach 1961, Holt and Omholt 1962, Brown and Barcus 1963, Gustafsson 1964, Rosenberg 1965, Barcus 1965).

Holt and Omholt (1962) made a detailed study of the correlation between the intensity of the auroral λ 5577 [OI] emission in the zenith measured by a photometer (5° field of view) and riometer measurements of absorption of cosmic radio noise on 27 MHz. The relation between the electron density N at a time t and the instantaneous electron production function q, is given by the equation of conservation

$$\frac{dN}{dt} = \frac{q}{1+\lambda} - \alpha N^2 , \tag{8.1}$$

where α is the effective recombination coefficient and λ the ratio between the number densities of negative ions and electrons. The electron production q is equal to the rate of ionization, and approximately proportional to the rates of emission of the first negative N_2^+ bands and the green oxygen line at 5577 Å (cf. Sect. s. 2.2 and 4.2.2).

In fact, the recombination process may be rather complicated, in particular in the D-region. Heavy cluster-ions may be involved, and a two-ion model proposed by Haug and Landmark (1970) would also lead to linear terms in the recombination. At high electron densities, such as during auroral conditions, this will be less important than under normal quiet conditions (cf. Folkestad and Armstrong 1970).

With the assumed proportionality between light output and rate of ionization, the brightness I observed from the ground at any time is

$$I = \int_1^2 \eta \, dh = k \int_1^2 q(h) \, dh , \tag{8.2}$$

where the integration is performed along the line of sight, η is the volume emission rate, and the constant k is the ratio between the rate of ionization and the optical emission. I, η and q are of course also functions of time t.

Under special assumptions, there exists a linear relation between the square root of the auroral luminosity and the cosmic radio noise absorption: if the energy spectrum and pitch angle distribution of the primary particles is constant in time, the ionization q at any height will be proportional to the brightness I observed from the ground (integrated emission, along the line of sight):

$$q(h,t) = c_1(h) I(t) . \tag{8.3}$$

The constant $c_1(h)$ depends on height and the particle energy spectrum. The absorption A as measured with a riometer in decibels can be expressed by the absorption coefficient K, which is proportional to the number density N of free electrons:

$$A(t) = \int_1^2 K(h,t)\mathrm{d}h = \int_1^2 c_2(h)N(h,t)\mathrm{d}h\,. \tag{8.4}$$

The integrals are performed throughout the ionosphere along the wave-path, with c_2 as a certain function of altitude and riometer frequency. Assuming quasi-equilibrium $\left(\frac{\mathrm{d}}{\mathrm{d}t} \approx 0\right)$, one obtains from Eq. (8.1):

$$N(h,t) = \alpha^{-\frac{1}{2}}(1+\lambda)^{-\frac{1}{2}} q^{\frac{1}{2}}(h,t) = c_3(h) q^{\frac{1}{2}}(h,t)\,. \tag{8.5}$$

Hence, by combination of Eqs. (8.3) to (8.5) and assuming identity of wave-path with line of sight, we get

$$I^{\frac{1}{2}}(t) = \left[\int_1^2 c_1^{\frac{1}{2}} c_2 c_3 \mathrm{d}h\right]^{-1} A(t) = c_4 A(t)\,. \tag{8.6}$$

The recombination coefficient, the negative ion to electron ratio, and the electron collision frequency (the latter is implicit in c_2) have been assumed to be independent of the primary electron flux. The ratio c_4 thus depends only on the electron energy spectrum. However, since Eq. (8.1) is an approximation only, c_4 must be considered as constant only to first approximation.

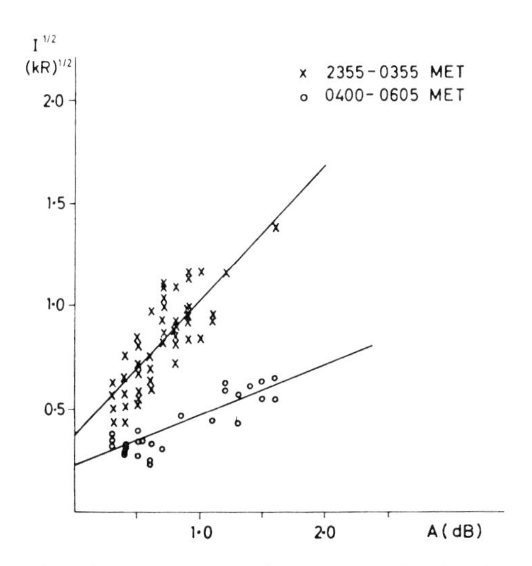

Fig. 8.1 Square root of auroral luminosity (λ 5577) versus absorption of 27.6 MHz radio waves, Tromsø (Johansen 1965, courtesy Pergamon Press)

It should be noted that, while the luminosity is largely proportional to the input of energy through the primary particles, the created electrons will be more efficient as absorbers the lower in the atmosphere they are produced. Absorption is thus more efficiently caused by high-energy electrons, and this is the reason why the factor c_4 in Eq. (8.6) depends on primary particle energy.

Holt and Omholt (1962) demonstrated that Eq. (8.6) was obeyed to a large extent, but that the factor c_4 varied from event to event. Similar results were obtained by Johansen (1965), using the same equipment. He found that the coefficient c_4 may remain constant for some hours and then change abruptly to a new value. Some of his results are shown in Fig. 8.1. Since the two instruments used differed greatly in view angle, he restricted the observations to cases when most of the sky was covered by fairly homogeneous aurora.

Johansen (1965) computed the coefficient c_4 theoretically for given primary electron energy spectra. With the present uncertainties in the basic parameters, Johansen estimated the uncertainty in the computed value of c_4 to be as much as a factor of two, but his observations clearly demonstrate the effect of the primary particle spectrum. For the two periods of the event shown in Fig. 8.1, and with exponential energy spectra, he found characteristic (e-folding) energies of 14 keV and 30 keV respectively, the harder energy spectrum being associated with the early morning period from 0400—0605 LT, giving more absorption relative to the optical intensity. Further, he used all his data to measure the characteristic (e-folding) energy as a function of time during the night, for widespread auroras. The variation was not significant.

Gustavson (1964) made a similar study and included studies of periods where appreciable changes occurred and the time variation in the electron density had to be taken into account. With a reasonable selection of basic data he found agreement between theoretical and observed time lags between peaks in auroral luminosity and absorption. He derived a recombination coefficient of about 2×10^{-6} cm^3 s^{-1}, which seems appropriate for a height of about 70 km. In the middle of break-up periods, the absorption may increase more than would be expected, possibly due to a hardening of the energy spectrum (Gustafsson 1969).

The relation between optical aurora and radio absorption has also been studied by Eather and Jacka (1966), Ansari (1964) and by Berkey (1968). Berkey (1968) used a narrow-beam riometer and was therefore able to make a detailed correlation study including all forms of aurora. His work represents a continuation of that of Ansari (1964), and is probably the most comprehensive so far. He divides the types of relation into three categories:

Category	I (5577)	Absorption
A	intense	intense
B	intense	small
C	small	intense

Category A was observed mostly near the midnight meridian, B in the evening sector and C in the morning sector. If a substorm is particularly intense, category A will also occur in the evening sector, accompanying a rapid westward surge.

Correspondingly, he finds that there is a systematic decrease of the constant c_4 in Eq. (8.6) during the course of the night. For λ 5577 c_4 decreases from about 5 to slightly more than $1\,\mathrm{kR}^{\frac{1}{2}}\,\mathrm{db}^{-1}$ between 1800 and 0600 local time. This indicates a hardening of the spectrum during the course of the night, as is also revealed by height and temperature variations (cf. Sect. 2.5 and 6.2—3). It differs from Johansen's (1965) result, which gave no significant changes; but these latter observations pertained to fairly widespread, quiet aurora. Berkey (1968) found significant changes in the ratio of luminosity to absorption only when either of the two quantities increased markedly. He also found that the spectrum was hardest on the equatorward side of the auroral oval, and during periods of pulsating aurora. Simultaneous observations of the $H\beta$ intensity indicated that the contribution to the D-layer ionization by protons was negligible.

For a long time it has been generally believed that pulsating aurora is associated with stronger absorption than usual (cf. Hargreaves 1969, Hultqvist 1965, 1969, Little 1967).

Campbell and Leinbach (1961) found a correlation of 0.7 between the λ 3914 fluctuations and auroral absorption, in support of earlier observations. Roldugin (1967) observed on occasions simultaneous pulsations in light emission and radio absorption with 10 to 60 s periods. Although it seems that both pulsations and strong absorption are typical post-midnight phenomena, doubt has recently been cast on the detailed correlation. Brekke (1969) investigated, for each hour during the night, the fractional occurrence of absorption above 0.5 db when optical pulsations were recorded and when pulsations were absent (Tromsø, geom. lat. 67°N). Using data from the four winters 1964 to 1968, he found no definite differences, and the two phenomena seemed to be statistically independent. The diurnal behaviour varied quite a bit from one winter to another, perhaps due to variations in general solar activity. The investigations confirmed that both absorption and pulsations are typical post-midnight phenomena, and this is probably the reason why a strong statistical dependence has often been implied.

With two phenomena which happen to occur at approximately the same time of the day, but are physically independent, the apparent correlation coefficient depends on the time resolution of the data.

An exception perhaps exists during the early morning hours, when fast pulsations are more abundant (Brekke 1969). Hajkowicz (1969) found that the morning absorption bays, probably caused by hard electron precipitation, were associated with pulsating aurora at the conjugate station.

8.3 E-Layer Ionization

Another radio phenomenon associated with optical aurora is the sporadic E-layer which is often observed when aurora occurs overhead (cf. e.g. King 1967, Bullen 1966). When the aurora is reasonably uniform over a sufficiently large area in the zenith, it will give rise to a fairly thick, smooth layer of enhanced ionization, giving a distinct reflection pattern on the iononogram. The height of the aurora is most commonly such that this auroral ionospheric layer coincides with the ordinary E-layer. The height and the electron density can usually be derived correctly from the ionogram, in contrast to the non-auroral sporadic E_s or the more erratic and sporadic E_a sometimes observed during aurora. Bullen (1966) suggested that this type of E-layer echo should be termed "Night E", and he used it to study the general auroral activity in the southern hemisphere during the years 1958 to 1963.

Cheney (1969) noted a systematic behaviour of the auroral ionization before and after short break-up periods (1—2 min) during which total absorption occurred. There was a rise before, and a fall afterwards, in the critical frequency, with a decrease and an increase in apparent height, respectively. Also, the percentage enhancement of f_0 E at break-up decreased systematically with increasing "quiet auroral" value (as determined some time before break-up).

Studies of the direct relation between the intensity of the aurora overhead and the electron density inferred from the ionograms have been made by Knecht (1956), Omholt (1955) and Oguti (1960) (cf. also Thomas 1967). Fig. 8.2 shows the simultaneous values of the emission of the first negative N_2^+ band λ 4278 and the maximum electron density overhead observed by Omholt (1955). Also included (large circles) are two points deduced from Knecht's (1956) data.

Adopting the standard recombination equation, Eq. (8.1), and neglecting time variations, the electron density N at any point is given by

$$N^2 = \frac{q}{\alpha(1+\lambda)}. \tag{8.7}$$

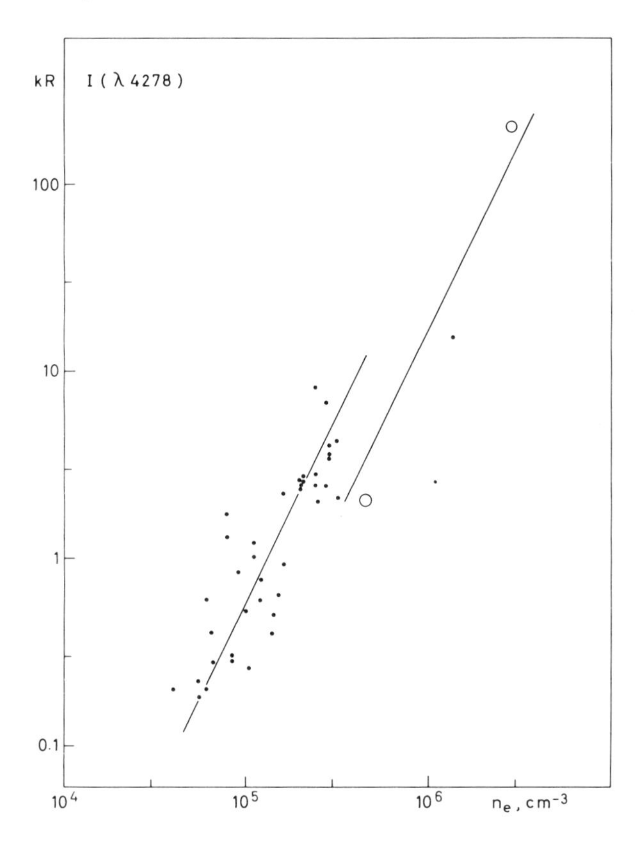

Fig. 8.2 Brightness of aurora in the zenith versus electron density of the "Night E" layer. ○ : From data by Knecht (1956)

According to the data given in Sect. 2.2, the total volume emission η of the first negative N_2^+ band λ 4278 should be about 75 times less than the total rate of ionization. If the luminosity distribution with height were constant, then η in the maximum point should always be proportional to brightness, with a proportionality constant h_0 which gives a measure of the thickness of the aurora. Hence (cf. also Eqs. 8.2 and 8.5),

$$N^2 = \frac{75\eta}{\alpha(1+\lambda)} = \frac{75I}{h_0\alpha(1+\lambda)} \tag{8.8}$$

where I is the total intensity of the first negative band λ 4278 measured in the zenith (photons cm^{-2} s^{-1}). As is seen from Fig. 8.2, there is a reasonable proportionality between N^2 and I, as shown by the straight lines. The scatter is to be expected, due to a certain dispersion in h_0 as well as in $\alpha(1+\lambda)$ because of the varying absolute height of the aurora.

From the data given in Sect. 2.5, 20 km is a reasonable estimate of the average value of h_0. From Fig. 8.2 values of 2×10^9 photons $cm^{-2}\,s^{-1}$ and 2×10^5 electrons cm^{-3} may be considered a typical pair of data. Inserting these values in Eq. (8.8) yields a value of about $2 \times 10^{-6}\,cm^3\,s^{-1}$ for $\alpha(1+\lambda)$, which is an order of magnitude larger than the estimates from a combination of optical, particle and electron density data obtained by rockets, and which were discussed in Sect. 2.2. The discrepancy may partly be due to calibration difficulties in the optical measurements, and is not necessarily real. The circles in Fig. 8.2, which are derived from Knecht's (1956) data, give an absolute value of the intensity corresponding to a given electron density which is about 3 times lower. If this is correct, $3-5 \times 10^{-7}\,cm^3\,s^{-1}$ seems to be the most probable value for the apparent recombination coefficient ($\alpha(1+\lambda)$) in aurora, at a height of 110—120 km.

8.4 Radio Aurora

Backscatter of radar waves is observed fairly regularly from regions where there is auroral activity. Reviews of this phenomenon, which is usually called radio aurora, have recently been given by Forsyth (1967) and Lange–Hesse (1967, 1969). There is agreement on the general explanation of the phenomenon: that high frequency radio waves are scattered by irregularities in the ionization at auroral heights, and that the appearance of such scattering is generally associated with aurora and magnetic storms. An important feature is that the scattering irregularities are aligned along the magnetic field, and that backscatter therefore occurs only from regions where the ray-path is approximately at right angles to the magnetic field (within about $\pm 10°$ at 40 MHz and about $\pm 4°$ at 100 MHz).

For several years there has been an almost continuous debate on the detailed correlation between the position of the scattering regions and the optical aurora. The reason for this uncertainty is largely due to the fact that scattering can only be located at low elevation angles, i.e. at distances of several hundred kilometers from the observing station, and that refraction suffered by the radio waves in the troposphere and the ionosphere may significantly affect the geometry (Unwin 1966, also Forsyth 1967). Because of these difficulties there have been few occasions when the appropriate measurements have been made.

Lange–Hesse and Czechowsky (1966) have observed a few cases when the locations of radio aurora and optical aurora were very close and partly coincident (cf. also Lange–Hesse 1967). Similarly, Kelly (1965) reported a close correlation between optical and radar measurements, and Bates *et al.* (1969) found a good correlation between the position and occurrence of precipitating electrons measured by a satellite, and the position of radar echoes on the night side. On the day side, part of the electron precipitation was not accompanied by radar echoes. In 14 instances when precise positions of optical aurora could be obtained through photometric triangulation, most of the echoes originated close to the narrow, discrete auroras used for triangulation measurements. They conclude that radio auroras are closely associated in time and space with optical auroras in the auroral oval, although there is not an exact correspondence. They find no correlation between radar echoes and precipitating protons.

In partial contrast to this and in agreement with his own earlier results, Gadsden (1967) found no direct association between the two phenomena. He correlated the intensity of optical aurora measured above one station with radar echoes observed from the same regions by another appropriately located station and found no correlation between echo occurrence and the measured intensity of the optical emissions at λ 5577, λ 3914 and λ 6300 Å. He concluded that the backscatter regions are not necessarily associated with the highest radiance observed, nor are the highest radiances always associated with backscatter regions. Unwin and Knox (1967) reached a similar conclusion, while McDiarmid and McNamara (1967) found that over a two-year period there was at least 50 percent correlation in space and time between optical aurora and radio aurora. They compared radar echoes from two widely separated radars with photographic records of the common volume. The variations in occurrence with seasons and with solar cycle are very similar for radio and optical aurora (Egeland 1962).

To explain the radio aurora it is necessary to assume small field-aligned irregularities in the electron density, a few meters in diameter and having a length of several meters. How such irregularities arise is still an open question. Although the optical measurements (cf. Sect. 2.6) have not yet given the lower limit for the structure in the primary particle stream, it is conceivable that the gyro-radii of the primary particles are too large to permit the fine structure inferred from radio observations. Also, the length of the ionized trails from primary particles of any single energy will be too long to fit the radio data. If the role of the aurora is to enhance the general electron density only, and irregularities are produced in this enhanced electron density, then only a limited correlation should exist. In this case there should be a correlation between the presence of radio and optical aurora, but the correlation between the intensities of the two phenomena may be very small.

Gadsden (1967) maintains that two-stream instability processes in the ionosphere may produce the observed effects without enhancement of the electron density. In that case, the correlation found should be that between optical aurora and the electric field producing the streams which are subject to the instability. McDiarmid and McNamara (1969) have evaluted a physical model of a radar aurora event. They assume two-stream instabilities caused by an electrojet flowing within a visual, east-west aligned aurora. There are considerable uncertainties involved in the calculations, but there is agreement with the particular set of observations upon which their model is based.

8.5 VLF Radio Emissions and Radio Noise

Radio emissions in the VLF region (i.e. below 30 kHz) have for a long time been associated with the presence of auroral phenomena, either directly with optical aurora or with other typical manifestations of aurora, such as radio absorption or magnetic disturbances (cf. Ungstrup 1967). Burton and Boardman (1933) were the first to recognize the correlation between VLF emissions and aurora. A statistical correlation between visual aurora and VLF hiss (band-limited thermal noise) has been found by several authors (Martin, Helliwell and Marks 1960, Jørgensen and Ungstrup 1962, Morozumi 1963, 1965). Harang, Larsen and Skogtvedt (1965) found strong correlation between aurora and VLF hiss. VLF emissions connected with auroral phenomena have recently been extensively studied by Harang (1968a, b) and by Holtet and Egeland (1969). Burton and Boardman (1933), Renard (1961) and Ungstrup (1966) have all observed VLF chorus (a multitude of overlapping, rising frequencies) during flaming aurora.

Hence there are numerous observations which show correlation of VLF emissions with optical aurora. According to Ungstrup (1967) VLF hiss in the 4—30 kHz region correlates with auroral arcs and bands in the pre-break-up and break-up phases of the aurora, while chorus is associated with flaming and pulsating aurora, and with the break-up phase in the auroral zone. Quantitative correlation between VLF emissions and optical aurora has not yet been established, but from rocket and ground-based data there is reason to suspect correlation in cases of weak aurora (Egeland, private comm.). With strong aurora, absorption may attenuate the VLF signal. Since both VLF and absorption of radio waves are correlated with optical aurora, it is not surprising that there is also a correlation between the two radio phenomena. Examples of detailed agreement of VLF emission and absorption have been presented (cf. Hargreaves 1969).

Radiation of radio noise in the MHz frequency region (30—3000 MHz) has been reported from time to time. Forsyth, Petrie and Currie (1949) detected the presence of noise pulses during auroral displays. Radio noise during auroral activity has been noted by Harz, Reid and Vogan (1957), Harz (1958), Seed (1958), Chivers and Wells (1959), and Egan and Peterson (1960). Others (Chapman and Currie 1953, Harang and Landmark 1954) have looked for radio noise during aurora with negative results, even during intense auroral displays. More recently Harang (1969) observed bursts of radio noise at 28 MHz at a sudden onset of radio absorption which apparently was due to aurora, while Eriksen and Harang (1969) observed radio noise on 225 MHz during great magnetic disturbances at the end of a PCA event.

It seems that radio noise in the MHz region is a rare phenomenon that appears only under special circumstances. No attempt has been made to correlate it quantitatively with optical aurora.

8.6 Polar Cap Absorption (PCA) Events

The PCA events are caused by high energy protons and alpha-particles precipitating approximately uniformly over large fractions of the polar caps, symmetrically around the magnetic poles, sometimes extending as far down as 50° geomagnetic latitude. The observed effect is a strong absorption of radio waves over the entire polar cap. The distinct polar glow aurora associated with this phenomenon was first reported by Sandford (1961, 1962) and confirmed by Weill (1962). A comprehensive discussion of the phenomenon has been given by Sandford (1967).

Polar glow auroras can be readily distinguished from other auroras inside the auroral oval only. The most striking optical characteristic is that the first negative N_2^+ bands are several times stronger than the green [OI] line at 5577 Å. The intensity of the N_2^+ band at 3914 Å exceeds that of the green line by a factor of 5 or more. The absolute intensity of the λ 3914 band in polar glow auroras ranges from 0.1 to 10 kR, the most typical values being 1 to 3 kR. Alcayde (1968) has made computations of the intensities of these emissions as a function of proton energies and the results are in reasonable agreement with observations.

The main reason for this particular intensity distribution is the height in the atmosphere where the glow occurs. Typical particle energies are of the order 10—100 MeV, and such protons and α-particles yield the main ionization and excitation effect at altitudes between 80 and 30 km. Although excitation of the forbidden oxygen lines also occurs at these heights through direct dissociation of molecular oxygen or dissociative recombination of molecular oxygen ions, there will be a pronounced collisional deactivation of the excited atoms, and almost all radiation of the λ 5577 line will be quenched below about 70 km (cf. Sect. 5.2). Argemi (1964) has reported occasional enhancements of the 6300 line during PCA events. This must be due to fluxes of low-energy particles and their direct connection with a PCA event seems questionable. Both the airglow and ordinary aurora will to a large degree contaminate observations of the forbidden oxygen emissions from PCA auroras.

Hydrogen emissions are not prominent in polar glow auroras for two reasons (cf. Chapt. 3): a) the $H\alpha$ and $H\beta$ yield per proton is not increased significantly when the primary energy exceeds 100 keV, hence the $H\alpha$ and $H\beta$ emission per unit energy input will be low for 100 MeV protons. b) When the charge-exchange reactions take place at low heights, deactivation due to re-ionization of excited H atoms becomes important. In fact, however, hydrogen emissions have been observed to be stronger than expected from combining the known energies of PCA protons with the intensity of the λ 3914 emission (cf. Sandford 1967). This points to simultaneous fluxes of low-energy protons, but again the difficulty arises with contamination by simultaneous appearance of other types of aurora.

Most other emissions observed in ordinary aurora seem to be present to some extent, but they have not been examined in detail. The relative intensities of all emissions from metastable levels are reduced compared to ordinary aurora, as would be expected from considerations on collisional deactivation at these heights (cf. Sect. 5.2).

References

Alcayde, D.: Ann. Geophys. **24**, 1031 (1968).
Ansari, Z. A.: J. Geophys. Res. **69**, 4493 (1964).
Argemi, L. H.: Ann. Geophys. **20**, 273 (1964).
Barcus, J. R.: J. Geophys. Res **70**, 2135 (1965).
Bates, H. F., Sharp, R. D., Belon, A. E., Boyd, J. S.: Planetary Space Sci. **17**, 83 (1969).
Berkey, F. T.: J. Geophys. Res. **73**, 319 (1968).
Brekke, A.: Thesis. University of Oslo 1969.
Brown, R. R., Barcus, J. R.: Arkiv. Geofysik **4**, 395 (1963).
Bullen, J. M.: J. Atmospheric Terrest. Phys. **28**, 879 (1966).
Burton, E. T., Boardman, E. M.: Proc. IRE **21**, 1476 (1933).
Campbell, W. H., Leinbach, H.: J. Geophys. Res. **66**, 25 (1961).
Chapman, R. P., Currie, B. W.: J. Geophys. Res. **58**, 363 (1953).
Cheney, B. J.: Australian J. Phys. **22**, 549 (1969).
Chivers, H. J. A., Wells, H. W.: J. Atmospheric Terrest. Phys. **17**, 13 (1959).
Eather, R. M., Jacka, F.: Australian J. Phys. **19**, 215 (1966).
Egan, R. D., Peterson, A. M.: J. Geophys. Res. **65**, 3830 (1960).
Egeland, A.: Arkiv Geofysik **4**, 103 (1962).

Eriksen, G., Harang, L.: Phys. Norvegica **4**, 1 (1969).
Folkestad, K., Armstrong, R.J.: J. Atmospheric Terrest. Phys. **32**, 409 (1970).
Forsyth, P.A., Petrie, W., Currie, B.W.: Nature, **164**, 453 (1949).
— In The Birkeland Symposium on Aurora and Magnetic Storms. (Ed. A. Egeland and J. Holtet), CNRS 1967.
Gadsden, M.: Planetary Space Sci. **15**, 693 (1967).
Gustafsson, G.: Planetary Space Sci. **12**, 195 (1964).
— Planetary Space Sci., **17**, 1961 (1969).
Hajkowicz, L. A.: J. Atmospheric Terrest. Phys. **31**, 1365 (1969).
Harang, L.: J. Atmospheric Terrest. Phys. **30**, 1143 (1968a).
— Planetary Space Sci. **16**, 1081 (1968b).
— Planetary Space Sci. **17**, 869 (1969).
— Landmark, B.: J. Atmospheric Terrest. Phys. **4**, 322 (1954).
— Larsen, R., Skogtvedt, J.: J. Atmospheric Terrest. Phys. **27**, 1147 (1965).
Hargreaves, J.K.: Proc. IEEE **57**, 1348 (1969).
Hartz, T.R.: Can. J. Phys. **36**, 677 (1958).
— Reid, G.C., Vogan, E.L.: Can. J. Phys. **34**, 728 (1956).
Haug, A., Landmark, B.: J. Atmospheric Terrest. Phys. **32**, 405 (1970).
Heppner, J.P., Byrne, E.C., Belon, A.E.: J. Geophys. Res. **57**, 121 (1952).
Holt, O., Omholt, A.: J. Atmospheric Terrest. Phys. **24**, 467 (1962).
Holtet, J., Egeland, A.: Phys. Norvegica **3**, 223 (1969).
Hultqvist, B.: In Introduction to Solar Terrestrical Relations. (Ed. J. Ortner and H. Maseland) Reidel Publ. Co. 1965.
— In Atmospheric Emissions. (Ed. B.M. McCormac and A. Omholt) Van Nostrand Reinhold Co. 1969.
Johansen, O.E.: Planetary Space Sci. **13**, 225 (1965).
Jørgensen, T.S., Ungstrup, E.: Nature **194**, 462 (1962).
Kelly, P.E.: Can. J. Phys. **43**, 1167 (1965).
King, G.A.M.: In Aurora and Airglow. (Ed. B.M. McCormac) Reinhold Publ. Co. 1967.
Knecht, R.W.: J. Geophys. Res. **61**, 59 (1956).
Lange-Hesse, G.: In Aurora and Airglow. (Ed. B.M. McCormac) Reinhold Publ. Co. 1967.
— In Atmospheric Emissions (Ed. B. M. McCormac and A. Omholt) Van Nostrand Reinhold. 1969.
— Czechowsky, P.: Arkiv elektr. Übertragung **20**, 365 (1966).
Little, C.G.: In Aurora and Airglow. (Ed. B.M. McCormac) Reinhold Publ. Co. 1967.
— Leinbach, H.: Proc. IRE. **46**, 334 (1958).
Martin, L.H., Helliwell, R.A., Marks, K.R.: Nature **187**, 751 (1960).
McDiarmid, D.R., McNamara, A.G.: Can. J. Phys. **45**, 3009 (1967).
— — Can. J. Phys. **47**, 1271 (1969).
Morozumi, H.N.: Natl. Acad. Sci. **73**, 16 (1963).
— Rep. Ionosph. Space Res. Japan **19**, 286 (1965).
Oguti, T.: Rept. Ionosph. Space Res. Japan **14**, 291 (1960).
Omholt, A.: J. Atmospheric Terrest. Phys. **7**, 73 (1955).
Renard, C.: C. R. Acad. Sci., Paris **252**, 1365 (1961).
Roldugin, V.K.: Geomagnetizm Aeronomiya (Engl. transl.) **7**, 454 (1967).
Rosenberg, T.J.: J. Atmospheric Terrest. Phys. **27**, 751 (1965).
Sandford, B.P.: Nature **190**, 245 (1961).
— J. Atmospheric Terrest. Phys. **24**, 155 (1962).
— Space Res. **7**, 836 (1967).

Seed, T.J.: J. Geophys. Res. **63**, 517 (1958).
Thomas, L.: In Aurora and Airglow. (Ed. B.M. McCormac) Reinhold Publ. Co. 1967.
Ungstrup, E.: J. Geophys. Res. **71**, 2395 (1966).
— In Aurora and Airglow. (Ed. B.M. McCormac) Reinhold Publ. Co. 1967.
Unwin, R.S.: J. Geophys. Res. **71**, 3677 (1966).
— Knox, F.B.: J. Atmospheric Terrest. Phys. **30**, 25 (1967).
Weill, G.: Compt. Rend. **254**, 3402 (1962).

Chapter 9

Auroral X-Rays

9.1 Production of X-Rays in Aurora

Soft radiation (as compared to cosmic rays) was detected at high latitudes from rocket flights in 1955 by Meredith *et al.* (1955, cf. also Van Allen 1957). The radiation was shown to be X-rays and thought to be produced as bremsstrahlung from fast electrons. The first direct association between X-rays and optical aurora was established by Winckler and associates in 1957 (Winckler and Peterson 1957, Winckler *et al.* 1958, 1959). At about the same time Anderson (1958) found that soft radiation events were associated with geomagnetic storms.

X-rays are produced by fast electrons as "bremsstrahlung" in their encounters with atoms and molecules. Coulomb forces between the fast electrons and the nuclei retard the electrons, and this gives rise to emission of X-rays with continuous spectral distribution, the maximum photon energy being equal to that of the electrons. The radiation process has been studied experimentally as well as theoretically (cf. Town Stephenson 1957, Chamberlain 1961). In addition, X-rays and ultraviolet radiation is produced by excitation of the inner-shell electrons in the atmospheric atoms and molecules.

For electrons in air with initial energy E_0, the total fraction of energy which is converted to continuous X-ray radiation is given approximately by:

$$f = 7 \times 10^{-9} \{E_0\}, \tag{9.1}$$

where $\{E_0\}$ is the numerical value of E_0 expressed in eV. The fraction which is converted to X-rays with energies between E and $E + \mathrm{d}E$ is approximately

$$\mathrm{d}f = 14 \times 10^{-9} \left(1 - \frac{E}{E_0}\right) \{\mathrm{d}E\}. \tag{9.2}$$

Hence, for a 100 keV electron about 70 eV of its energy is converted into X-rays, whereas for a 10 keV electron only 0.7 eV is converted. The directional distribution of these X-rays is not isotropic; the actual polar diagram depends on energy.

While the production of X-rays (per unit input energy) is proportional to the specific energy of the electrons, the light production is almost independent of the energy. Therefore, the relation between auroral luminosity and X-ray emission is a function of the energy spectrum of the primary particles. For an exponential energy distribution (exp $[-E/E_1]$) the ratio between total X-ray radiation and total light yield is proportional to the characteristic energy E_1. Because of this, we must expect the relation between X-ray emission and light to vary strongly, the X-ray emission being dominated by the high-energy tail of the electron energy spectrum. Thus, if the interpretation of fast intensity variations in optical aurora suggested on the basis of temperature measurements (cf. Sect. 6.2) is correct (*i.e.* that they are mainly due to increase in specific particle energy and not to increase in particle flux), then such intensity variations should appear greatly enhanced when recorded by X-ray emission.

9.2 Observations

X-rays from auroras have mainly been measured by means of balloon-borne equipment at altitudes of about 30—40 km. Since X-rays are heavily absorbed in the atmosphere, and the more so the lower the specific energy, even this technique is restricted to X-rays with energies of about 20 keV or above. The various types of X-ray emission events as observed from balloons are roughly described in Table 9.1, which is based on data given in two review papers by Kremser (1967, 1969). We shall not go into details on the X-ray emission as such, but rather concentrate on the relation between X-rays and light (X-rays will be discussed in detail in a later volume of *Physics and Chemistry in Space*).

The X-rays are not radiated isotropically relative to the direction of the electron path, and they are scattered in the atmosphere. Therefore, the quantitative relation between the X-ray spectrum and intensity measured at any particular point and the properties of the aurora above is rather complicated. Comparison is easiest when the auroral form is broad and diffuse, and in such cases variations in the ratio between X-ray intensity and light intensity may fairly directly reflect variations in the energy spectrum of the primary electrons.

In agreement with these considerations, the observations carried out show a fairly loose correspondence between X-rays and light (Winckler *et al.* 1958, Anderson and Enemark 1960, Bhavsar 1961, Anderson 1962, Anderson and De Witt 1963, Barcus 1965, Rosenberg 1965). Anderson and De Witt (1963) found a fairly good correlation between X-ray intensity and auroral luminosity, estimated from all-sky camera

Table 9.1 *Characteristics of X-ray events (from data by Kremser* 1967, 1969*)*

Event type	Time of occurrence	Precipitation region	Typical energy spectrum (e-folding energy)
Long lasting, smoothly varying events	After midnight until noon, sometimes also early afternoon	Auroral zone 100° or more in longitude	Systematic diurnal variation. 30 keV at 01 L.T. 45 keV at 16 L.T.
Impulsive bursts	Late evening and night, mainly around local midnight		10—15 keV
Slow pulsations (several minutes)	Daytime	Up to more than 1000 km	30—35 keV
Intermediate pulsations (80—100 s)	Daytime	More than 300 km	30 keV
Fast pulsations	Morning	About 100 km	15 keV
Microbursts (halfwidth below 0.2 s)	Morning and around noon	About 100 km	25 keV

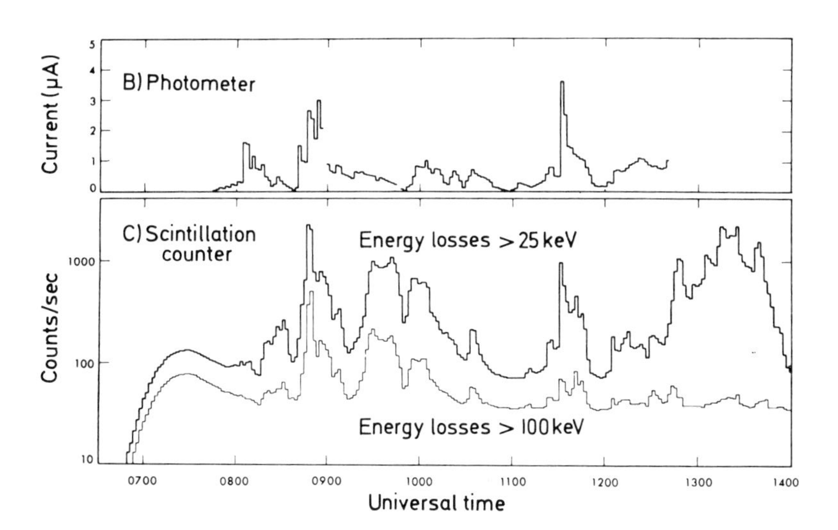

Fig. 9.1 Simultaneous optical (2950—3200 Å) and X-ray emissions from aurora, observed by balloon-borne detectors (Rosenberg 1965, courtesy Pergamon Press)

pictures, during widespread, diffuse aurora (glow). Fig. 9.1 shows a record by Rosenberg (1965), which demonstrates the qualitative relationship often found during events with discrete aurora.

More detailed studies of variations in X-ray intensities and auroral luminosity were made by Rosenberg *et al.* (1971) (cf. also Rosenberg *et al.* 1967). They used balloon-borne X-ray detectors and photometers. Although the gross variation of X-ray intensity and optical intensity often differed greatly, they found a close correlation in pulsations with a few seconds period, often with coherent pulsations in the two intensities. Bearing in mind that the geometrical factors and view angle are very different, this correspondence is most remarkable. An example of the records is shown in Fig. 9.2. The data indicate that even if the energy spectrum of the primary particles changes slowly, the modulation mechanism responsible for the pulsations works in similar fashion on electrons of all energies. Pettersen *et al.* (1968) compared the power spectra of the two sets of simultaneous pulsations, and concluded that the modulation mechanism was situated far out in the magnetosphere. This conclusion was also reached by Brown *et al.* (1965 a, b) from their studies of the conjugacy of X-ray pulsations. The correlation between pulsations in X-rays and auroral light is also discussed in Sect. 7.7.

X-rays appear, at least on some occasions, as a conjugate phenomenon. (Anderson *et al.* 1962, cf. also Brown *et al.* 1963) detected X-rays simultane-

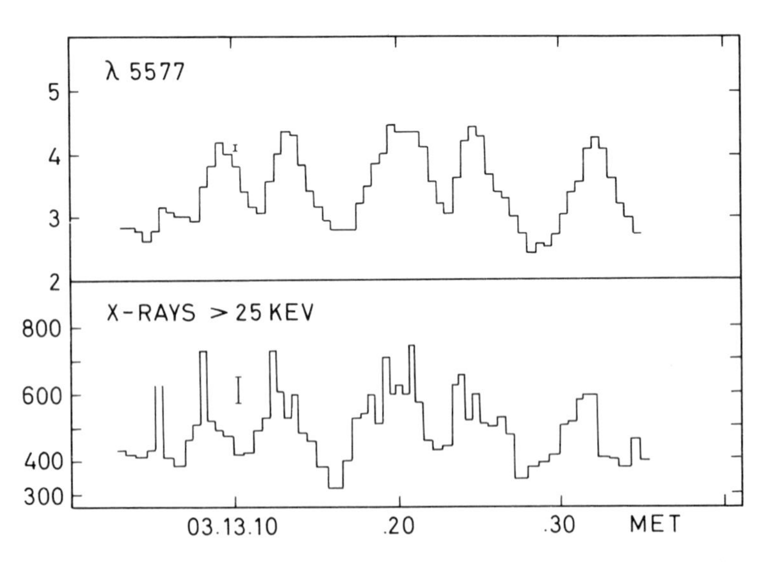

Fig. 9.2 Simultaneous pulsations in optical (λ 5577) and X-ray emissions from aurora, observed by balloon-borne detectors. Units are kR and counts per second, respectively (From data by Rosenberg *et al.* 1970)

ously in conjugate areas in the north and in the south. The time variations and counting rates were very similar in balloon-borne instruments at nearly conjugate points (differing about 15° in geomagnetic longitude and 3° in latitude). Brown *et al*; (1965 a, b) found conjugacy in slow time variations, pulsations and microbursts, but in the latter two types of events similarity in the phenomena in the north and in the south occurred only when the balloons were at more closely conjugated points. Such observations, together with optical ones, could shed light not only on the conjugacy of each one of the two phenomena, optical and X-ray auroras, but also allow a detailed study of differences and variations in characteristic features in the energy spectrum of the primary particles at conjugate points.

Barcus (1965) and Rosenberg (1965) conclude that precipitation of electrons at high and low energies is due to essentially different, but not infrequently coupled mechanisms. The correlation of the two phenomena depends on the degree of coupling. However, there is no particular evidence in the X-ray observations that the nature of the acceleration process is different at high and at low energies. Qualitatively, the X-ray observations are in agreement with other evidence from optical data as well as from direct particle measurements, that the energies and energy spectrum of the primary particles may vary considerably from one aurora to another, as well as during the same auroral event. X-ray observations are unique because they provide sensitive, stationary measurements of the high energy tail of the electron spectrum.

References

Anderson, K. A.: Phys. Rev. **111,** 1397 (1958).
— J. Physical. Soc. Japan **A-1, 17,** 237 (1962).
— Anger, C. D., Brown, R. R., Evans, D. S.: J. Geophys. Res. **67,** 4076 (1962).
— DeWitt, R.: J. Geophys. Res. **68,** 2669 (1963).
— Enemark, D. C.: J. Geophys. Res. **65,** 3521 (1960).
Barcus, J. R.: J. Geophys. Res. **70,** 2135 (1965).
Bhavsar, P. D.: J. Geophys. Res. **66,** 679 (1961).
Brown, R. R., Anderson, K. A., Anger, C. D., Evans, D. S.: J. Geophys. Res. **68,** 2677 (1963).
— Barcus, J. R., Parsons, N. R.: J. Geophys. Res. **70,** 2579 (1965a).
— J. Geophys. Res. **70,** 2599 (1965b).
Chamberlain, J. W.: Physics of the Aurora and Airglow. Academic Press 1961.
Kremser, G.: In Aurora and Airglow. (Ed. B. M. McCormac) Reinhold Publ. Co. 1967.
— in Atmospheric Emissions. (Ed. B. M. McCormac and A. Omholt) Van Nostrand Reinhold Co. 1969.
Meredith, L. H., Gottlieb, M. B., Van Allen, J. A.: Phys. Rev. **97,** 201 (1955).

Pettersen, H., Kvifte, G. J., Bjordal, J.: In The Birkeland Symposium on Aurora and Magnetic Storms. (Ed. J. Holtet and A. Egeland.) CNRS 1968.
Rosenberg, T. J.: J. Atmospheric Terrest. Phys. **27,** 751 (1965).
— Bjordal, J., Kvifte, G. J.: J. Geophys. Res. **72,** 3504 (1967).
— — Trefall, H., Kvifte, G. J., Omholt, A., Egeland, A.: J. Geophys. Res. **76,** 122 (1971).
Town Stephenson, S.: Handbuch der Physik **30,** 337. Berlin–Göttingen–Heidelberg: Springer 1957.
Van Allen, J. A.: Proc. Natl. Acad. Sci. **43,** 57 (1957).
Winckler, J. R., Peterson, L.: Phys. Rev. **108,** 903 (1957).
— — Arnoldy, R., Hoffman, R.: Phys. Rev. **110,** 1221 (1958).
— — J. Geophys. Res. **64,** 597 (1959).

Subject Index

Typesetting and printing: Zechnersche Buchdruckerei Speyer
Bookbinding: Konrad Triltsch, Würzburg